T0296490

LONDON MATHEMATICAL SOCIETY STUDENT TEXTS

Managing editor: Professor C.M. Series, Mathematics Institute
University of Warwick, Coventry CV4 7AL, United Kingdom

London Mathematical Society Student Texts 49

An Introduction to *K*-Theory for *C**-Algebras

M. Rørdam,
University of Copenhagen

F. Larsen
University of Odense

N. Laustsen
University of Leeds

CAMBRIDGE
UNIVERSITY PRESS

CAMBRIDGE UNIVERSITY PRESS
Cambridge, New York, Melbourne, Madrid, Cape Town, Singapore, São Paulo

Cambridge University Press
The Edinburgh Building, Cambridge CB2 8RU, UK

Published in the United States of America by Cambridge University Press, New York

www.cambridge.org
Information on this title: www.cambridge.org/9780521783347

First published 2000

A catalogue record for this publication is available from the British Library

ISBN 978-0-521-78334-7 hardback
ISBN 978-0-521-78944-8 paperback

Transferred to digital printing 2007

Contents

Preface

About K-theory

K-theory was developed by Atiyah and Hirzebruch in the 1960s based on work of Grothendieck in algebraic geometry. It was introduced as a tool in C^*-algebra theory in the early 1970s through some specific applications described below. Very briefly, K-theory (for C^*-algebras) is a pair of functors, called K_0 and K_1, that to each C^*-algebra A associate two Abelian groups $K_0(A)$ and $K_1(A)$. The group $K_0(A)$ is given an ordering that (in special cases) makes it an ordered Abelian group. There are powerful machines, some of which are described in this book, making it possible to calculate the K-theory of a great many C^*-algebras. K-theory contains much information about the individual C^*-algebras — one can learn about the structure of a given C^*-algebra by knowing its K-theory, and one can distinguish two C^*-algebras from each other by distinguishing their K-theories. For certain classes of C^*-algebras, K-theory is actually a complete invariant. K-theory is also a natural home for index theory.

Two applications demonstrated the importance of K-theory to C^*-algebras. George Elliott showed in the early 1970s (in a work published in 1976, [18]) that AF-algebras (the so-called "approximately finite dimensional" C^*-algebras; see Chapter 7 for a precise definition) are classified by their ordered K_0-groups. (The K_1-group of an AF-algebra is always zero.) As a consequence, all information about an AF-algebra is contained in its ordered K_0-group. This result indicated the possibility of classifying a more general class of C^*-algebras by their K-theory.

Another important application of K-theory to C^*-algebras was Pimsner and Voiculescu's proof in 1982, [34], of the fact that $C^*_{\mathrm{red}}(F_2)$, the reduced C^*-algebra of the free group of two generators, has no projections other than 0 and 1. Kadison had, at a time when it was not known that there exists a simple unital C^*-algebra with no projections other than 0 and 1, conjectured that $C^*_{\mathrm{red}}(F_2)$ would be such an example. It was shown by Powers in 1975 in [35] that $C^*_{\mathrm{red}}(F_2)$ actually is a simple C^*-algebra, and, as mentioned, Pimsner and Voiculescu then showed that this C^*-algebra has no non-trivial projections by calculating its K-theory. Blackadar had a couple of years before that found another example of a simple unital projectionless C^*-algebra [2].

A landmark for the use of K-theory in C^*-algebra theory, and for the use of K-theory for C^*-algebras in topology, was Brown, Douglas, and Fillmore's development in the 1970s of K-homology (a dual theory to K-theory) via extensions of C^*-algebras [8] and [9]. This theory was generalized by Kasparov in his KK-theory that encompasses K-theory and K-homology [27].

Today K-theory is an active research area, and a much used tool for the study of C^*-algebras. One current line of research concentrates on generalizing Elliott's classification theorem for AF-algebras to a much broader class of C^*-algebras; see [19]. Another active branch of research seeks to prove the conjecture by Baum and Connes on the K-theory of the C^*-algebra $C^*_{\mathrm{red}}(G)$ of an arbitrary group G in a way that generalizes Pimsner and Voiculescu's result about $C^*_{\mathrm{red}}(F_2)$. Connes has in his book Noncommutative Geometry, [11], described how K-theory is useful in the understanding of a big mathematical landscape that contains geometry, physics, C^*-algebras, and algebraic topology among many other subjects!

Besides actually being useful — in the mathematical sense of the word — K-theory is fun to study because of the way it mixes ideas from the different branches of mathematics where it has its roots.

The aim of this book is not to present all the new mathematics that involves K-theory for C^*-algebras, but to give an elementary and, we hope, easy-to-read introduction to the subject.

About the book

This book first saw the light of day as a set of handwritten lecture notes to a graduate course on K-theory for C^*-algebras at Odense University, Denmark, in the spring of 1995, given by the first named author and with the two other authors among the participants. The handwritten notes were TEX'ed and rewritten in the fall of 1995 as a joint project among the three authors. The K-theory course has since then been given once more in Odense (in the fall of 1997) and once again in Copenhagen (in the fall of 1998). Besides, a number of students have taken a reading course based on these notes. We have in this way received substantial feedback from the many students who have been subjected to the notes and, as a result, the notes have continuously been improved. We started writing this book in winter 1998/99 in order to make the work that over the years has been put into the notes available to a larger group of readers.

The book is intended as a text for a one-semester first or second year graduate course, or as a text for a reading course for students at that level. As such, the text does not take the reader very far into the vast world of K-theory. The most basic properties of K_0 and K_1 are covered, Bott periodicity is proved and the six-term exact sequence is derived. There is a chapter on inductive limits and continuity of K-theory, and Elliott's classification of AF-algebras is proved. In the last chapter of the book, the theory developed is used to show that each pair of countable Abelian groups arises as the K-groups of a C^*-algebra.

An effort has been made to make the text as self-contained as possible. Chapter 1 contains an overview, mostly without proofs, of what the reader should know, or should learn, about C^*-algebras. The theory in the book is illustrated with examples that can be found in the exercises and in the text.

The subject of this book is treated in several other textbooks, most notably in Bruce Blackadar's book, [3]. Other books treating K-theory for operator algebras include Niels Erik Wegge-Olsen's detailed treatment, [40], Gerard Murphy's book, [29], and the recent books by Ken Davidson, [15], and by Peter Fillmore, [20]. Our treatment is indebted to these books, in particular to Bruce Blackadar's book.

We thank Hans Jørgen Munkholm for sharing with us his point of view on K-theory. We thank also George Elliott for many valuable comments. Last, but not least, we thank those who have read and commented on (earlier versions of) this text. Thanks are especially due to Piotr Dzierzynski, Jacob Hjelmborg, Johan Kustermans, Mikkel Møller Larsen, Franz Lehner, Jesper Mygind, Agata Przybyszewska, Rolf Dyre Svegstrup, and Steen Thorbjørnsen.

Sections, examples, and paragraphs in the book marked with an asterisk * contain digressions which the reader can omit or postpone without losing the logic of the overall exposition. Two possible shorter routes through the book are

(1) Chapters 1–4 and Chapters 8–12, or
(2) Chapters 1–7.

Chapter 7 and Chapter 13 can be omitted or postponed.

The book has a home page

$$\texttt{http://www.math.ku.dk/}\sim\texttt{rordam/K-theory.html}$$

that will contain a list of corrections to the book. Readers are strongly encouraged to report on any mistakes they may have found in the book (see the home page for address information). We also welcome suggestions of how to make the book better.

Chapter 1

C^*-Algebra Theory

This chapter contains some basic facts about C^*-algebras that the reader is assumed to be (or become) familiar with. There are very few proofs given in this chapter, and the reader is referred to other sources, for example Murphy's book [29], for details.

1.1 C^*-algebras and *-homomorphisms

Definition 1.1.1. A C^*-*algebra* A is an algebra over $\mathbb{C}$ with a *norm* $a \mapsto \|a\|$ and an *involution* $a \mapsto a^*$, $a \in A$, such that A is complete with respect to the norm, and such that $\|ab\| \leqslant \|a\| \, \|b\|$ and $\|a^*a\| = \|a\|^2$ for every a, b in A.

The axioms for a C^*-algebra A above imply that the involution is isometric, i.e., $\|a\| = \|a^*\|$ for every a in A.

A C^*-algebra A is called *unital* if it has a multiplicative identity, which will be denoted by 1 or 1_A. A *-homomorphism $\varphi \colon A \to B$ between C^*-algebras A and B is a linear and multiplicative map which satisfies $\varphi(a^*) = \varphi(a)^*$ for all a in A. If A and B are unital and $\varphi(1_A) = 1_B$, then φ is called *unital* (or *unit preserving*). A C^*-algebra is said to be *separable* if it contains a countable dense subset.

1.1.2 Sub-C^*-algebras and sub-*-algebras. A non-empty subset B of a C^*-algebra A is called a *sub-*-algebra* of A if it is a *-algebra with the operations given on A, that is, if it is closed under the algebraic operations:

addition	$A \times A \to A,$	$(a, b) \mapsto a + b,$
multiplication	$A \times A \to A,$	$(a, b) \mapsto ab,$
adjoint	$A \to A,$	$a \mapsto a^*,$
scalar multiplication	$\mathbb{C} \times A \to A,$	$(\alpha, a) \mapsto \alpha a.$

A *sub-C^*-algebra* of A is a non-empty subset of A which is a C^*-algebra with respect to the operations given on A. Hence, a non-empty subset B of a C^*-algebra A is a sub-C^*-algebra if and only if it is norm-closed and closed under the four algebraic operations listed above.

The norm-closure of a sub-*-algebra of a C^*-algebra is a sub-C^*-algebra. This follows from the fact that the four algebraic operations above are continuous.

Let A be a C^*-algebra, and let F be a subset of A. The sub-C^*-algebra of A generated by F, denoted by $C^*(F)$, is the smallest sub-C^*-algebra of A that contains F. In other words, $C^*(F)$ is the intersection of all sub-C^*-algebras of A that contain F. The C^*-algebra $C^*(F)$ can be concretely described as follows. For each natural number n put

$$W_n = \{x_1 x_2 \cdots x_n : x_j \in F \cup F^*\},$$

where $F^* = \{x^* \mid x \in F\}$, and put $W = \bigcup_{n=1}^{\infty} W_n$. The set W is the set of all *words* in $F \cup F^*$, and W_n is the set of words of length n. Using that $W = W^*$ and that W is closed under multiplication, we see that the linear span of W is a sub-*-algebra of A. It follows that

$$C^*(F) = \overline{\operatorname{span} W}.$$

We write $C^*(a_1, a_2, \ldots, a_n)$ instead of $C^*(\{a_1, a_2, \ldots, a_n\})$, when $a_1, a_2, \ldots, a_n$ are elements in A.

Theorem 1.1.3 (Gelfand–Naimark). *For each C^*-algebra A there exist a Hilbert space H and an isometric *-homomorphism φ from A into $B(H)$, the algebra of all bounded linear operators on H. In other words, every C^*-algebra is isomorphic to a sub-C^*-algebra of $B(H)$. If A is separable, then H can be chosen to be a separable Hilbert space.*

A proof can be found in [29, Theorem 3.4.1]. The Hilbert space H is obtained by viewing A as a vector space, equipping it with a suitable inner product (coming from a positive, linear functional on A), and forming the completion of A with respect to this inner product.

1.1.4 Ideals and quotients. By an ideal in a C^*-algebra we shall always understand a closed, two-sided ideal (unless otherwise stated). Every ideal is automatically self-adjoint, and thereby a sub-C^*-algebra (see [29, 3.1.3]).

Assume that I is an ideal in a C^*-algebra A. The quotient of A by I is

$$A/I = \{a + I : a \in A\}, \quad \|a + I\| = \inf\{\|a + x\| : x \in I\}, \quad \pi(a) = a + I.$$

In this way A/I becomes a C^*-algebra, $\pi \colon A \to A/I$ is a *-homomorphism, called the *quotient mapping*, and $I = \text{Ker}(\pi)$ (see [29, 3.1.4]).

Let $\varphi \colon A \to B$ be a *-homomorphism. Then, automatically, $\|\varphi(a)\| \leqslant \|a\|$ for all a in A, and φ is injective if and only if φ is isometric (see Exercise 1.8 or [29, 3.1.5]). The kernel, $\text{Ker}(\varphi)$, of φ is an ideal in A, and the image, $\text{Im}(\varphi) = \varphi(A)$, of φ is a sub-C^*-algebra of B (see [29, 3.1.6]). By the first isomorphism theorem there is one (and only one) *-homomorphism $\varphi_0 \colon A/\text{Ker}(\varphi) \to B$ such that the diagram

commutes, i.e., $\varphi_0 \circ \pi = \varphi$. Moreover, φ_0 is injective.

A C^*-algebra A is called *simple* if the only ideals in A are the two trivial ideals 0 and A.

1.1.5 Short exact sequences. A (finite or infinite) sequence of C^*-algebras and *-homomorphisms

$$\cdots \longrightarrow A_n \xrightarrow{\varphi_n} A_{n+1} \xrightarrow{\varphi_{n+1}} A_{n+2} \longrightarrow \cdots$$

is said to be *exact* if $\text{Im}(\varphi_n) = \text{Ker}(\varphi_{n+1})$ for all n. An exact sequence of the form

$$(1.1) \qquad\qquad 0 \longrightarrow I \xrightarrow{\varphi} A \xrightarrow{\psi} B \longrightarrow 0$$

is called *short exact*.

If I is an ideal in A, then

$$0 \longrightarrow I \xrightarrow{\iota} A \xrightarrow{\pi} A/I \longrightarrow 0$$

is a short exact sequence, where ι is the inclusion mapping and π is the quotient mapping. Conversely, given (1.1), then $\varphi(I)$ is an ideal in A, the C^*-algebra B is isomorphic to $A/\varphi(I)$, and we have a commutative diagram

$$
\begin{array}{ccccccccc}
0 & \longrightarrow & I & \xrightarrow{\varphi} & A & \xrightarrow{\psi} & B & \longrightarrow & 0 \\
& & \varphi\downarrow\cong & & \| & & \downarrow\cong & & \\
0 & \longrightarrow & \varphi(I) & \xhookrightarrow{\iota} & A & \xrightarrow{\pi} & A/\varphi(I) & \longrightarrow & 0.
\end{array}
$$

If in (1.1) there is a *-homomorphism $\lambda\colon B \to A$ such that $\psi \circ \lambda = \mathrm{id}_B$, then λ is called a *lift* of ψ and (1.1) is said to be *split exact*. Not all short exact sequences are split exact, see Exercise 1.2.

The *direct sum* $A \oplus B$ of two C^*-algebras A and B is the C^*-algebra of all pairs (a, b), with $a \in A$ and $b \in B$, equipped with entry-wise defined algebraic operations and the norm

$$\|(a, b)\| = \max\{\|a\|, \|b\|\}.$$

Define $\iota_A\colon A \to A \oplus B$, $\iota_B\colon B \to A \oplus B$, $\pi_A\colon A \oplus B \to A$, and $\pi_B\colon A \oplus B \to B$ by $\iota_A(a) = (a, 0)$, $\iota_B(b) = (0, b)$, $\pi_A(a, b) = a$, and $\pi_B(a, b) = b$. Then

$$0 \longrightarrow A \xrightarrow{\iota_A} A \oplus B \underset{\iota_B}{\overset{\pi_B}{\rightleftarrows}} B \longrightarrow 0$$

is a split exact sequence (with lift ι_B). Not all split exact sequences are direct sums, see Exercise 1.3 (vi) and Exercise 1.1.

1.1.6 Adjoining a unit. To every C^*-algebra A one can associate a unique unital C^*-algebra $\widetilde{A}$ that contains A as an ideal and with the property that $\widetilde{A}/A$ is isomorphic to $\mathbb{C}$. See Exercise 1.3 and [29, 2.1.6] for details.

Let $\pi\colon \widetilde{A} \to \mathbb{C}$ be the quotient mapping, and let $\lambda\colon \mathbb{C} \to \widetilde{A}$ be defined by $\lambda(\alpha) = \alpha 1_{\widetilde{A}}$. Then

(1.2) $$0 \longrightarrow A \xrightarrow{\iota} \widetilde{A} \underset{\lambda}{\overset{\pi}{\rightleftarrows}} \mathbb{C} \longrightarrow 0$$

is a split exact sequence, and

$$\widetilde{A} = \{a + \alpha 1_{\widetilde{A}} : \ a \in A, \ \alpha \in \mathbb{C}\}.$$

The C^*-algebra $\widetilde{A}$ is called the *unitization* of A, or A with a unit adjoined.

Adjoining a unit is functorial in the sense that if $\varphi \colon A \to B$ is a *-homomorphism, then there is one and only one *-homomorphism $\widetilde{\varphi} \colon \widetilde{A} \to \widetilde{B}$ making the diagram

$$\begin{array}{ccccccccc}
0 & \longrightarrow & A & \longrightarrow & \widetilde{A} & \longrightarrow & \mathbb{C} & \longrightarrow & 0 \\
& & \varphi \downarrow & & \widetilde{\varphi} \downarrow & & \| & & \\
0 & \longrightarrow & B & \longrightarrow & \widetilde{B} & \longrightarrow & \mathbb{C} & \longrightarrow & 0
\end{array}$$

commutative. The *-homomorphism $\widetilde{\varphi}$ is given by $\widetilde{\varphi}(a + \alpha 1_{\widetilde{A}}) = \varphi(a) + \alpha 1_{\widetilde{B}}$, where a belongs to A and α is a complex number. Observe that $\widetilde{\varphi}$ is unit preserving.

If A is contained in a unital C^*-algebra B whose unit 1_B does not belong to A, then $\widetilde{A}$ is equal (or isomorphic) to the sub-C^*-algebra $A + \mathbb{C}{\cdot}1_B$ of B. (See Exercise 1.3.)

If A has a unit 1_A, and if $1_{\widetilde{A}}$ as above is the unit in $\widetilde{A}$, then $f = 1_{\widetilde{A}} - 1_A$ is a projection in $\widetilde{A}$, and

$$\widetilde{A} = \{a + \alpha f : a \in A, \ \alpha \in \mathbb{C}\}.$$

The map $A \oplus \mathbb{C} \to \widetilde{A}$ given by $(a, \alpha) \mapsto a + \alpha f$ is a *-isomorphism, and $\widetilde{A}$ is isomorphic to $A \oplus \mathbb{C}$. If A is not unital, then $\widetilde{A}$ is not isomorphic to $A \oplus \mathbb{C}$ because $\widetilde{A}$ is unital and $A \oplus \mathbb{C}$ is not.

1.2 Spectral theory

1.2.1 The spectrum. Let A be a unital C^*-algebra and let a be an element in A. The *spectrum* of a is the set of complex numbers λ such that $a - \lambda{\cdot}1$ is not invertible, and it is denoted by $\mathrm{sp}(a)$. The *spectral radius*, $r(a)$, of a is

$$r(a) = \sup\{|\lambda| : \lambda \in \mathrm{sp}(a)\}.$$

The spectrum $\mathrm{sp}(a)$ is a closed subset of $\mathbb{C}$, and the spectral radius satisfies $r(a) \leqslant \|a\|$ (see [29, Lemma 1.2.4] for these two facts). It follows that $\mathrm{sp}(a)$ is a compact subset of the complex plane. A more refined argument, using complex analysis, shows that the spectrum is non-empty, the sequence

$\{\|a^n\|^{1/n}\}_{n=1}^{\infty}$ is decreasing, and

$$(1.3) \qquad\qquad r(a) = \lim_{n\to\infty} \|a^n\|^{1/n}.$$

If A is not unital, then embed A in its unitization $\widetilde{A}$ (see Paragraph 1.1.6) and let $\mathrm{sp}(a)$ be the spectrum of a viewed as an element in $\widetilde{A}$. If A is non-unital, then $0 \in \mathrm{sp}(a)$ for all a in A.

An element a in A is called

- *self-adjoint* if $a = a^*$,

- *normal* if $aa^* = a^*a$,

- *positive* if a is normal and $\mathrm{sp}(a) \subseteq \mathbb{R}^+$ (with the convention that 0 belongs to $\mathbb{R}^+$).

The set of positive elements in A is denoted by A^+. It is an elementary fact that every self-adjoint element has spectrum contained in $\mathbb{R}$. It is a much deeper fact that an element a in A is positive if and only if $a = x^*x$ for some x in A. For a normal element a, equation (1.3) reduces to $r(a) = \|a\|$.

The spectrum of an element a in a C^*-algebra A depends, a priori, on the ambient C^*-algebra A. However, if A is a *unital* C^*-algebra, B is a sub-C^*-algebra of A that contains the unit of A, and a is an element in B, then the spectrum of a relative to B is equal to the spectrum of a relative to A. When A is non-unital, or when A is unital, but the unit of A does not belong to B, then the spectrum, relative to A, of an element a in B consists of 0 and of all numbers in the spectrum of a relative to B. These claims follow from the fact that if a is an invertible element in a unital C^*-algebra A, then the inverse of a belongs to $C^*(a)$, the smallest sub-C^*-algebra of A containing a, see Exercise 1.6.

1.2.2 States. A linear map $\rho\colon A \to \mathbb{C}$ is called a (linear) functional. Let

$$\|\rho\| = \sup\{|\rho(a)| : a \in A,\ \|a\| \leqslant 1\}$$

denote its operator norm. A linear functional ρ is continuous if and only if $\|\rho\| < \infty$. If $\rho(a) \geqslant 0$ for every positive element a in A, then ρ is said to be positive. A *state* ρ on a *unital* C^*-algebra A is a positive linear functional with $\rho(1) = 1$, or, equivalently, with $\|\rho\| = 1$. The set of states on a unital C^*-algebra separates points in the sense that if a is an element in A such that $\rho(a) = 0$ for every state ρ on A, then $a = 0$.

The next theorem (and the additions to it) characterizing commutative C^*-algebras has the interpretation that a general (non-commutative) C^*-algebra can be viewed as a "non-commutative topological space", and that the study of C^*-algebras is the study of "non-commutative topology". K-theory for C^*-algebras is a branch of operator algebras where this analogy holds particularly well.

Theorem 1.2.3 (Gelfand). *Every Abelian C^*-algebra is isometrically *-isomorphic to the C^*-algebra $C_0(X)$ for some locally compact Hausdorff space X.*

Recall that $C_0(X)$ is the C^*-algebra of all continuous functions $f\colon X \to \mathbb{C}$ "vanishing at infinity": for each $\varepsilon > 0$ there is a compact subset K of X such that $|f(x)| \leqslant \varepsilon$ for all x in $X \setminus K$. The norm on $C_0(X)$ is the supremum norm. If X is compact, then $C_0(X)$ is equal to $C(X)$, the set of all continuous functions $f\colon X \to \mathbb{C}$.

In addition to Gelfand's theorem we have the following.

(i) $C_0(X)$ is unital if and only if X is compact.

(ii) $C_0(X)$ is separable if and only if X is separable.

(iii) X and Y are homeomorphic if and only if $C_0(X)$ and $C_0(Y)$ are isomorphic.

(iv) Each continuous function $g\colon Y \to X$ induces a *-homomorphism $\varphi\colon C_0(X) \to C_0(Y)$ by $\varphi(f) = f \circ g$, and, conversely, for every *-homomorphism $\varphi\colon C_0(X) \to C_0(Y)$ there is a continuous function $\widehat{\varphi}\colon Y \to X$, such that $\varphi(f) = f \circ \widehat{\varphi}$.

(v) There is a bijective correspondence between open subsets of X and ideals in $C_0(X)$. The ideal corresponding to the open subset U of X is the set of all f in $C_0(X)$ that vanish on the complement, U^c, of U, and this ideal is isomorphic to $C_0(U)$. (The isomorphism is given by restriction, and one shows that this map is surjective by invoking the Stone–Weierstrass theorem.) The *-homomorphism $C_0(X) \to C_0(U^c)$ given by restriction $f \mapsto f|_{U^c}$ is surjective (by the Stone–Weierstrass theorem), and so we get a short exact sequence:

$$0 \longrightarrow C_0(U) \longrightarrow C_0(X) \longrightarrow C_0(U^c) \longrightarrow 0.$$

The map $C_0(U) \to C_0(X)$ is given by extending a function f in $C_0(U)$ to X by giving it the value 0 on U^c.

1.2.4 The continuous function calculus. Let A be a unital C^*-algebra. To each normal element a in A there is one and only one *-isomorphism

$$C(\mathrm{sp}(a)) \to C^*(a, 1) \subseteq A, \qquad f \mapsto f(a),$$

which maps ι to a, where ι in $C(\mathrm{sp}(a))$ is given by $\iota(z) = z$ for all z in $\mathrm{sp}(a)$. This unique *-isomorphism has the properties that its assignment of $f(a)$ agrees with the usual definition when f is a polynomial, and that $\tau(a) = a^*$ when $\tau\colon \mathbb{C} \to \mathbb{C}$ is the function $\tau(z) = \bar{z}$.

The *spectral mapping theorem* says that $\mathrm{sp}(f(a)) = f(\mathrm{sp}(a))$ for every normal element a and every continuous function f on $\mathrm{sp}(a)$.

If $\varphi\colon A \to B$ is a unital *-homomorphism between the C^*-algebras A and B, and if a is a normal element in A, then $\mathrm{sp}(\varphi(a)) \subseteq \mathrm{sp}(a)$, and $f(\varphi(a)) = \varphi(f(a))$ for every f in $C(\mathrm{sp}(a))$.

If a is a normal element in a non-unital C^*-algebra A, then $f(a)$ is, a priori, defined to be an element in the unitization $\tilde{A}$ of A. With $\pi\colon \tilde{A} \to \mathbb{C}$ the quotient mapping, since $\pi(f(a)) = f(\pi(a)) = f(0)$ when a belongs to A, we see that $f(a)$ belongs to A precisely when $f(0) = 0$.

In Chapter 2 we shall need the following result.

Lemma 1.2.5. *Let K be a non-empty compact subset of $\mathbb{R}$, and let $f\colon K \to \mathbb{C}$ be a continuous function. Let A be a unital C^*-algebra, and let Ω_K be the set of self-adjoint elements in A with spectrum contained in K. The (induced) function*

$$f\colon \Omega_K \to A, \qquad a \mapsto f(a),$$

is continuous.

Proof. The map $A \to A$ given by $a \mapsto a^n$ is continuous for every integer $n \geqslant 0$ (because multiplication is continuous). It follows that every complex polynomial f induces a continuous map $A \to A$ given by $a \mapsto f(a)$.

Now, let $f\colon K \to \mathbb{C}$ be any continuous function, let a be an element in Ω_K, and let $\varepsilon > 0$. By the Stone–Weierstrass theorem there is a complex polynomial g such that $|f(z) - g(z)| \leqslant \varepsilon/3$ for every z in K. Find $\delta > 0$ such that $\|g(a) - g(b)\| \leqslant \varepsilon/3$ whenever b is an element in A with $\|a - b\| \leqslant \delta$. Since

$$\|f(c) - g(c)\| = \|(f - g)(c)\| = \sup\{|(f - g)(z)| : z \in \mathrm{sp}(c)\} \leqslant \varepsilon/3$$

for all c in Ω_K, it follows that $\|f(a) - f(b)\| \leqslant \varepsilon$ for all b in Ω_K with $\|a-b\| \leqslant \delta$. $\qquad \square$

A more elaborate argument shows that the conclusion of Lemma 1.2.5 holds in the more general situation where K is any non-empty (compact or not) subset of $\mathbb{C}$, and where Ω_K is the set of all normal elements in A with spectrum contained in K.

1.3 Matrix algebras

For each C^*-algebra A and for each natural number n, let $M_n(A)$ be the set of all $n \times n$ matrices

$$\begin{pmatrix} a_{11} & a_{12} & \cdots & a_{1n} \\ a_{21} & a_{22} & \cdots & a_{2n} \\ \vdots & \vdots & \ddots & \vdots \\ a_{n1} & a_{n2} & \cdots & a_{nn} \end{pmatrix},$$

where each matrix entry a_{ij} belongs to A. Equip $M_n(A)$ with the usual entry-wise vector space operations and matrix multiplication, and set

$$\begin{pmatrix} a_{11} & a_{12} & \cdots & a_{1n} \\ a_{21} & a_{22} & \cdots & a_{2n} \\ \vdots & \vdots & \ddots & \vdots \\ a_{n1} & a_{n2} & \cdots & a_{nn} \end{pmatrix}^* = \begin{pmatrix} a_{11}^* & a_{21}^* & \cdots & a_{n1}^* \\ a_{12}^* & a_{22}^* & \cdots & a_{n2}^* \\ \vdots & \vdots & \ddots & \vdots \\ a_{1n}^* & a_{2n}^* & \cdots & a_{nn}^* \end{pmatrix}.$$

To define a C^*-norm on $M_n(A)$, choose a Hilbert space H and an injective *-homomorphism $\varphi \colon A \to B(H)$. Let $\varphi_n \colon M_n(A) \to B(H^n)$ be given by

$$\varphi_n \begin{pmatrix} a_{11} & \cdots & a_{1n} \\ \vdots & \ddots & \vdots \\ a_{n1} & \cdots & a_{nn} \end{pmatrix} \begin{pmatrix} \xi_1 \\ \vdots \\ \xi_n \end{pmatrix} = \begin{pmatrix} \varphi(a_{11})\xi_1 + \cdots + \varphi(a_{1n})\xi_n \\ \vdots \\ \varphi(a_{n1})\xi_1 + \cdots + \varphi(a_{nn})\xi_n \end{pmatrix}, \quad \xi_j \in H.$$

Define a norm on $M_n(A)$ by $\|a\| = \|\varphi_n(a)\|$ for a in $M_n(A)$. With these operations, $M_n(A)$ becomes a C^*-algebra, the norm is independent of the choice of representation φ provided that it is injective, and

$$(1.4) \qquad \max_{i,j}\{\|a_{ij}\|\} \leqslant \left\| \begin{pmatrix} a_{11} & \cdots & a_{1n} \\ \vdots & \ddots & \vdots \\ a_{n1} & \cdots & a_{nn} \end{pmatrix} \right\| \leqslant \sum_{i,j} \|a_{ij}\|.$$

A proof of the the inequality above is outlined in Exercise 1.13. It shows for example that a function $f \colon X \to M_n(A)$ is continuous if and only if each entry function $f_{ij}\colon X \to A$ is continuous. We shall occasionally use the abbreviation

$$\operatorname{diag}(a_1, a_2, \ldots, a_n) = \begin{pmatrix} a_1 & 0 & \cdots & 0 \\ 0 & a_2 & \cdots & 0 \\ \vdots & \vdots & \ddots & \vdots \\ 0 & 0 & \cdots & a_n \end{pmatrix}$$

for a *diagonal matrix*, where $a_1, a_2, \ldots, a_n$ are in A.

Forming matrix algebras has the functorial property that if A and B are C^*-algebras and if $\varphi\colon A \to B$ is a *-homomorphism, then the map $\varphi_n\colon M_n(A) \to M_n(B)$ given by

$$(1.5) \qquad \varphi_n \begin{pmatrix} a_{11} & \cdots & a_{1n} \\ \vdots & \ddots & \vdots \\ a_{n1} & \cdots & a_{nn} \end{pmatrix} = \begin{pmatrix} \varphi(a_{11}) & \cdots & \varphi(a_{1n}) \\ \vdots & \ddots & \vdots \\ \varphi(a_{n1}) & \cdots & \varphi(a_{nn}) \end{pmatrix}$$

is a *-homomorphism for each natural number n. We shall often omit the index n and write φ instead of φ_n.

1.4   Exercises

Exercise 1.1. You are given a short exact sequence

$$0 \longrightarrow A \overset{\varphi}{\longrightarrow} E \overset{\psi}{\longrightarrow} B \longrightarrow 0.$$

In the notation of Paragraph 1.1.5, show that there is a *-isomorphism $\theta\colon E \to A \oplus B$ that makes the diagram

$$
\begin{array}{ccccccccc}
0 & \longrightarrow & A & \overset{\varphi}{\longrightarrow} & E & \overset{\psi}{\longrightarrow} & B & \longrightarrow & 0 \\
 & & \| & & \downarrow{\scriptstyle\theta} & & \| & & \\
0 & \longrightarrow & A & \underset{\iota_A}{\longrightarrow} & A \oplus B & \underset{\pi_B}{\longrightarrow} & B & \longrightarrow & 0
\end{array}
$$

commutative if and only if there is a *-homomorphism $\nu\colon E \to A$ such that $\nu \circ \varphi = \operatorname{id}_A$.

Exercise 1.2. Show that

$$0 \longrightarrow C_0((0,1)) \overset{\iota}{\longrightarrow} C([0,1]) \overset{\psi}{\longrightarrow} \mathbb{C} \oplus \mathbb{C} \longrightarrow 0$$

is a short exact sequence, where $\psi(f) = (f(0), f(1))$, and show that the sequence does not split.

Exercise 1.3. The purpose of this exercise is to check the claims of Paragraph 1.1.6. Let A be a C^*-algebra (with or without a unit). Set
$$\widetilde{A} = \{(a, \alpha) : a \in A, \ \alpha \in \mathbb{C}\},$$
define addition and scalar multiplication coordinate-wise on $\widetilde{A}$, and define multiplication and involution by
$$(a, \alpha) \cdot (b, \beta) = (ab + \beta a + \alpha b, \alpha\beta), \qquad (a, \alpha)^* = (a^*, \overline{\alpha}).$$
Let $\iota \colon A \to \widetilde{A}$ and $\pi \colon \widetilde{A} \to \mathbb{C}$ be given by $\iota(a) = (a, 0)$ and $\pi(a, \alpha) = \alpha$, respectively.

(i) Justify that $\widetilde{A}$ is a $*$-algebra. Show that the element $(0, 1)$ in $\widetilde{A}$ is a unit for $\widetilde{A}$, that ι is an injective $*$-homomorphism, and that π is a surjective $*$-homomorphism.

We shall hereafter denote the unit element $(0, 1)$ by $1_{\widetilde{A}}$ or by 1.

Suppressing ι we obtain that A is contained in $\widetilde{A}$ and that
$$\widetilde{A} = \{a + \alpha \cdot 1 : a \in A, \ \alpha \in \mathbb{C}\}.$$
Let $\|\cdot\|_A$ be the norm on A. For each x in $\widetilde{A}$, set
$$\||x|\|_{\widetilde{A}} = \sup\{\|ax\|_A : a \in A, \ \|a\|_A \leqslant 1\},$$
and define $\|x\|_{\widetilde{A}}$ to be the maximum of the two numbers $\||x|\|_{\widetilde{A}}$ and $|\pi(x)|$.

(ii) Show that $\|a\|_{\widetilde{A}} = \|a\|_A$ for all a in A.

(iii) Let x be in $\widetilde{A}$. Show that $x = 0$ if $\|x\|_{\widetilde{A}} = 0$.

(iv) Show that $\widetilde{A}$ is a unital C^*-algebra. More specifically, show that $\|\cdot\|_{\widetilde{A}}$ is a norm, that $\|xy\|_{\widetilde{A}} \leqslant \|x\|_{\widetilde{A}}\|y\|_{\widetilde{A}}$, and that $\|x^*x\|_{\widetilde{A}} = \|x\|_{\widetilde{A}}^2$ for all x, y in A, and show that $\widetilde{A}$ is complete with respect to $\|\cdot\|_{\widetilde{A}}$.

(v) Show that the sequence (1.2)
$$0 \longrightarrow A \overset{\iota}{\longrightarrow} \widetilde{A} \underset{\lambda}{\overset{\pi}{\rightleftarrows}} \mathbb{C} \longrightarrow 0$$
is split exact. Conclude that A is an ideal in $\widetilde{A}$.

(vi) Show that $\widetilde{A}$ is isomorphic to $A \oplus \mathbb{C}$ (as C^*-algebras) if and only if A is unital.

Exercise 1.4. An element p in a C^*-algebra is a *projection* if $p = p^* = p^2$, and an element u in a unital C^*-algebra is said to be *unitary* if $uu^* = u^*u = 1$.

(i) Let p be a projection. Show that $\mathrm{sp}(p) \subseteq \{0,1\}$.

(ii) Suppose that p is a normal element and that $\mathrm{sp}(p) \subseteq \{0,1\}$. Show that p is a projection.

(iii) Let u be a unitary element. Show that $\mathrm{sp}(u) \subseteq \mathbb{T}$, where

$$\mathbb{T} = \{z \in \mathbb{C} : |z| = 1\}.$$

(iv) Suppose that u is a normal element and that $\mathrm{sp}(u) \subseteq \mathbb{T}$. Show that u is unitary.

Exercise 1.5. Let A be a unital C^*-algebra. Suppose that A contains a normal element a whose spectrum is not connected. Show that A contains a non-trivial projection, i.e., a projection other than 0 and 1_A.

Show that if A is a (possibly non-unital) C^*-algebra with a normal element a with non-connected spectrum, then A contains a non-zero projection.

Exercise 1.6. Let A be a unital C^*-algebra, and let a be an element in A.

(i) Show that a is invertible if and only if aa^* and a^*a are invertible, and show that $a^{-1} = (a^*a)^{-1}a^* = a^*(aa^*)^{-1}$ in this case.

(ii) Suppose that b is an invertible and normal element in A. Show that $b^{-1} = f(b)$ for some function f in $C(\mathrm{sp}(b))$.

(iii) Show that if a is invertible, then a^{-1} belongs to $C^*(a)$, the smallest sub-C^*-algebra of A that contains a. [Hint: Use (i) and (ii).]

Exercise 1.7. Show that each element x in a C^*-algebra A has a unique decomposition $x = a + ib$, where a, b are self-adjoint elements in A. [Hint: Calculate $(a + ib)^*$.]

Exercise 1.8. Let A and B be C^*-algebras, and let $\varphi \colon A \to B$ be a unital *-homomorphism.

(i) Show that $\mathrm{sp}(\varphi(a)) \subseteq \mathrm{sp}(a)$ for each a in A; and show that these two sets are equal if φ is injective.

(ii) Show that $\|\varphi(a)\| \leqslant \|a\|$ for all a in A; and show that $\|\varphi(a)\| = \|a\|$ for all a in A if φ is injective. [Hint: Use (i) applied to a^*a together with the fact that $r(c) = \|c\|$ for each *normal* element c, see Paragraph 1.2.1.]

Exercise 1.9. Let X be a compact Hausdorff space, and let μ be a Borel measure on X. Let H be the Hilbert space $L^2(X, \mu)$. For every f in $C(X)$ and for every ξ in H, define $M_f\xi \colon X \to \mathbb{C}$ by $(M_f\xi)(x) = f(x)\xi(x)$.

(i) Show that $\|M_f\xi\|_2 \leqslant \|f\|_\infty \|\xi\|_2$, where $\|\cdot\|_\infty$ is the supremum norm on $C(X)$. Conclude that $M_f\xi$ belongs to H, that M_f belongs to $B(H)$, and that $\|M_f\| \leqslant \|f\|_\infty$.

Define $\pi \colon C(X) \to B(H)$ by $\pi(f) = M_f$.

(ii) Show that π is a *-homomorphism.

(iii) Show that π is isometric (and hence injective) if the measure μ has the following property: every non-empty open subset U of X contains a Borel subset E satisfying $0 < \mu(E) < \infty$.

(iv) Suppose that X is separable. Construct a Borel measure μ on X that satisfies the condition in (iii).

Exercise 1.10. Let X be a (not necessarily locally compact or Hausdorff) topological space. Say that $C_0(X)$ *separates points in* X if for each pair of distinct points x_1, x_2 in X there is f in $C_0(X)$ such that $f(x_1)$ and $f(x_2)$ are distinct and non-zero. Show that $C_0(X)$ separates points in X if and only if X is a locally compact Hausdorff space.

Exercise 1.11. Let A be a C^*-algebra, and let X be a locally compact Hausdorff space. Consider the vector space $C_0(X, A)$ of all continuous functions $f \colon X \to A$ such that for every $\varepsilon > 0$ there is a compact subset K of X such that $\|f(x)\| \leqslant \varepsilon$ for every x in $X \setminus K$. For f, g in $C_0(X, A)$, define fg and f^* in $C_0(X, A)$ by $(fg)(x) = f(x)g(x)$ and $f^*(x) = f(x)^*$ for all x in X, and set

$$\|f\| = \sup\{\|f(x)\| : x \in X\}.$$

Show that $C_0(X, A)$ is a C^*-algebra with these operations.

Exercise 1.12. Let A be a unital C^*-algebra, and let x be an element in $M_2(A)$. Show that

$$x \begin{pmatrix} 1 & 0 \\ 0 & 0 \end{pmatrix} = \begin{pmatrix} 1 & 0 \\ 0 & 0 \end{pmatrix} x$$

if and only if $x = \operatorname{diag}(a, b)$ for some a, b in A. Show a and b are unitary if x is unitary.

Exercise 1.13. Prove the inequality (1.4). [Hint: Given $a = (a_{ij})$ in $M_n(A)$, let $a^{(ij)}$ in $M_n(A)$ be the matrix whose (i,j)th entry is a_{ij} and whose other entries are zero, so that $a = \sum_{i,j} a^{(ij)}$. Show that $\|a^{(ij)}\| = \|a_{ij}\|$, for example by representing the C^*-algebra A on a Hilbert space, and use this to prove (1.4).]

Exercise 1.14. Let A be a unital C^*-algebra and consider the *upper triangular matrix*

$$a = \begin{pmatrix} a_{11} & a_{12} & a_{13} & \cdots & a_{1n} \\ 0 & a_{22} & a_{23} & \cdots & a_{2n} \\ 0 & 0 & a_{33} & \cdots & a_{3n} \\ \vdots & \vdots & \vdots & \ddots & \vdots \\ 0 & 0 & 0 & \cdots & a_{nn} \end{pmatrix}$$

in $M_n(A)$. Show that a is invertible in $M_n(A)$ if and only if all diagonal entries $a_{11}, a_{22}, \ldots, a_{nn}$ are invertible in A. [Hint: Solve the equation $ab = 1$ where a is as above and where b is an unknown upper triangular matrix.]

Exercise 1.15. Show that

$$\left\| \begin{pmatrix} a & 0 \\ 0 & b \end{pmatrix} \right\| = \max\{\|a\|, \|b\|\}.$$

for each pair of elements a, b in a C^*-algebra A. [Hint: Show that the map $(a,b) \mapsto \mathrm{diag}(a,b)$ defines an injective — and hence isometric — *-homomorphism from $A \oplus A$ to $M_2(A)$.]

Chapter 2

Projections and Unitary Elements

The K-theory of a C^*-algebra is defined in terms of equivalence classes of its projections and equivalence classes of its unitary elements — possibly after adjoining a unit and forming matrix algebras. We shall in this chapter derive the facts needed about projections and unitary elements with emphasis on the equivalence relation defined by homotopy and also — for projections — Murray–von Neumann equivalence and unitary equivalence.

2.1 Homotopy classes of unitary elements

2.1.1 Homotopy. Let X be a topological space. Say that two points a, b in X are *homotopic* in X, written $a \sim_h b$ in X, if there is a continuous function $v \colon [0,1] \to X$ such that $v(0) = a$ and $v(1) = b$. The relation $\sim_h$ is an equivalence relation on X. The continuous function v above is called a *continuous path* from a to b, and it is often denoted by $t \mapsto v(t)$ or $t \mapsto v_t$, with or without specifying explicitly that t belongs to the interval $[0,1]$.

Needless to say, the reference to the space X is crucial. For example, any two elements a, b in a C^*-algebra A are homotopic in A. Indeed, take the continuous path $t \mapsto (1-t)a + tb$. But, as we shall see, two projections in A need not be homotopic in the set of all projections in A. We shall nevertheless sometimes omit the reference to the space X and just write $a \sim_h b$ instead of $a \sim_h b$ in X, when it is clear from the context in which space the homotopy should be realized.

Definition 2.1.2. Let A be a unital C^*-algebra. Recall that an element u in A is unitary if $uu^* = 1 = u^*u$. Denote the group of unitary elements in A by $\mathcal{U}(A)$. From Paragraph 2.1.1 we have the equivalence relation $\sim_h$ on $\mathcal{U}(A)$. Let $\mathcal{U}_0(A)$ be the set of all u in $\mathcal{U}(A)$ such that $u \sim_h 1$ in $\mathcal{U}(A)$.

If u_1, v_1, u_2, v_2 are unitary elements in a C^*-algebra A with $u_1 \sim_h v_1$ and $u_2 \sim_h v_2$, then $u_1 u_2 \sim_h v_1 v_2$. Indeed, find continuous paths $t \mapsto w_j(t)$ in $\mathcal{U}(A)$ from u_j to v_j for $j = 1, 2$. Then $t \mapsto w_1(t) w_2(t)$ is a continuous path in $\mathcal{U}(A)$ from $u_1 u_2$ to $v_1 v_2$.

Recall in (ii) below that the spectrum of a unitary element is contained in the unit circle $\mathbb{T}$, see Exercise 1.4.

Lemma 2.1.3. *Let A be a unital C^*-algebra.*

(i) *For each self-adjoint element h in A, $\exp(ih)$ belongs to $\mathcal{U}_0(A)$.*

(ii) *If u is a unitary element in A with $\mathrm{sp}(u) \neq \mathbb{T}$, then u belongs to $\mathcal{U}_0(A)$.*

(iii) *If u, v are unitary elements in A with $\|u - v\| < 2$, then $u \sim_h v$.*

Proof. (i). It follows from the continuous function calculus (see Paragraph 1.2.4) that if $f \colon \mathbb{R} \to \mathbb{T}$ is continuous and if h is a self-adjoint element in A, then $f(h)^* = \bar{f}(h) = f^{-1}(h)$. This shows that $f(h)$ is unitary, and in particular that $\exp(ih)$ is unitary. For each t in $[0,1]$ define $f_t \colon \mathrm{sp}(h) \to \mathbb{T}$ by $f_t(x) = \exp(itx)$. Since $t \mapsto f_t$ is continuous, so is the path $t \mapsto f_t(h)$ in $\mathcal{U}(A)$. Therefore $\exp(ih) = f_1(h) \sim_h f_0(h) = 1$.

(ii). If $\mathrm{sp}(u) \neq \mathbb{T}$, then $\exp(i\theta)$ does not belong to $\mathrm{sp}(u)$ for some real number θ. Let φ be the real function on $\mathrm{sp}(u)$ defined by $\varphi(\exp(it)) = t$, where t belongs to the open interval $(\theta, \theta + 2\pi)$. Then φ is continuous, and $z = \exp(i\varphi(z))$ for all z in $\mathrm{sp}(u)$. Set $h = \varphi(u)$. It follows that h is a self-adjoint element in A with $u = \exp(ih)$, and so u belongs to $\mathcal{U}_0(A)$ by (i).

(iii). If $\|u - v\| < 2$, then $\|v^*u - 1\| = \|v^*(u - v)\| < 2$ because each unitary element has norm 1. It follows that -2 is not in $\mathrm{sp}(v^*u - 1)$, see Paragraph 1.2.1. This shows that -1 does not belong to $\mathrm{sp}(v^*u)$. We can then use (ii) to conclude that $v^*u \sim_h 1$, and hence that $u \sim_h v$ as desired. $\square$

Each unitary element in $M_n(\mathbb{C})$ has finite spectrum and therefore belongs to $\mathcal{U}_0(M_n(\mathbb{C}))$ by Lemma 2.1.3 (ii). This proves the following Corollary.

Corollary 2.1.4. *The unitary group in $M_n(\mathbb{C})$ is connected, in other words,* $\mathcal{U}_0(M_n(\mathbb{C})) = \mathcal{U}(M_n(\mathbb{C}))$.

Lemma 2.1.5 (Whitehead). *Let A be a unital C^*-algebra and let u, v be unitary elements in A. Then*

$$\begin{pmatrix} u & 0 \\ 0 & v \end{pmatrix} \sim_h \begin{pmatrix} uv & 0 \\ 0 & 1 \end{pmatrix} \sim_h \begin{pmatrix} vu & 0 \\ 0 & 1 \end{pmatrix} \sim_h \begin{pmatrix} v & 0 \\ 0 & u \end{pmatrix} \quad in \ \ \mathcal{U}(M_2(A)).$$

It follows in particular that

$$\begin{pmatrix} u & 0 \\ 0 & u^* \end{pmatrix} \sim_h \begin{pmatrix} 1 & 0 \\ 0 & 1 \end{pmatrix} \quad in \ \ \mathcal{U}(M_2(A)).$$

Proof. From Lemma 2.1.3 (ii) we get

$$\begin{pmatrix} 0 & 1 \\ 1 & 0 \end{pmatrix} \sim_h \begin{pmatrix} 1 & 0 \\ 0 & 1 \end{pmatrix}.$$

Hence

$$\begin{pmatrix} u & 0 \\ 0 & v \end{pmatrix} = \begin{pmatrix} u & 0 \\ 0 & 1 \end{pmatrix} \begin{pmatrix} 0 & 1 \\ 1 & 0 \end{pmatrix} \begin{pmatrix} v & 0 \\ 0 & 1 \end{pmatrix} \begin{pmatrix} 0 & 1 \\ 1 & 0 \end{pmatrix} \sim_h \begin{pmatrix} uv & 0 \\ 0 & 1 \end{pmatrix}.$$

The other claims follow in a similar way.   □

Proposition 2.1.6. *Let A be a unital C^*-algebra.*

(i) *$\mathcal{U}_0(A)$ is a normal subgroup of $\mathcal{U}(A)$.*

(ii) *$\mathcal{U}_0(A)$ is open and closed relative to $\mathcal{U}(A)$.*

(iii) *An element u in A belongs to $\mathcal{U}_0(A)$ if and only if*

$$u = \exp(ih_1) \cdot \exp(ih_2) \cdots \exp(ih_n)$$

for some natural number n and some self-adjoint elements $h_1, h_2, \ldots, h_n$ in A.

Proof. To show (i) note first that $\mathcal{U}_0(A)$ is closed under multiplication by the comment below Definition 2.1.2. We must also show that if u belongs to $\mathcal{U}_0(A)$, then so do u^{-1} and vuv^* for every v in $\mathcal{U}(A)$. Let $t \mapsto w_t$ be a continuous path in $\mathcal{U}(A)$ from 1 to u. Then $t \mapsto w_t^{-1}$ and $t \mapsto vw_tv^*$ are continuous paths in $\mathcal{U}(A)$ from 1 to u^{-1} and from 1 to vuv^*, respectively. This proves (i).

Let G be the set of elements in A of the form

$$\exp(ih_1) \cdot \exp(ih_2) \cdots \exp(ih_n),$$

where n is a natural number and $h_1, h_2, \ldots, h_n$ are self-adjoint elements in A. It follows from (i) and from Lemma 2.1.3 (i) that G is contained in $\mathcal{U}_0(A)$. Since $\exp(ih)^{-1} = \exp(-ih)$ for every self-adjoint element h, we see that G is a group.

Take u in $\mathcal{U}(A)$ and v in G with $\|u - v\| < 2$. Then $\|1 - uv^*\| = \|u - v\| < 2$, and by Lemma 2.1.3 (iii) and its proof, $uv^* = \exp(ih)$ for some self-adjoint element h in A. Hence $u = \exp(ih)v$, and u therefore belongs to G. This shows that G is open relative to $\mathcal{U}(A)$.

The complement $\mathcal{U}(A) \setminus G$ is a disjoint union of cosets of the form Gu with u in $\mathcal{U}(A)$. Each of these cosets is homeomorphic to G and therefore open (relative to $\mathcal{U}(A)$). Consequently G is closed in $\mathcal{U}(A)$.

In conclusion, G is a non-empty subset of $\mathcal{U}_0(A)$, G is open and closed in $\mathcal{U}(A)$, and $\mathcal{U}_0(A)$ is connected. This entails that $\mathcal{U}_0(A) = G$ and that (ii) and (iii) hold. $\square$

Lemma 2.1.7. *Let A and B be unital C^*-algebras, and let $\varphi \colon A \to B$ be a surjective (and hence unit preserving) $*$-homomorphism.*

(i) $\varphi(\mathcal{U}_0(A)) = \mathcal{U}_0(B)$.

(ii) *For each u in $\mathcal{U}(B)$ there exists v in $\mathcal{U}_0(M_2(A))$ such that*

$$\varphi_2(v) = \begin{pmatrix} u & 0 \\ 0 & u^* \end{pmatrix},$$

where $\varphi_2 \colon M_2(A) \to M_2(B)$ is the map induced by φ.

(iii) *If u is a unitary element in B, and if there is a unitary element v in A such that $u \sim_h \varphi(v)$, then u belongs to $\varphi(\mathcal{U}(A))$, i.e., u lifts to a unitary element in A.*

Proof. (i). A unital $*$-homomorphism is continuous and maps unitary elements to unitary elements. Therefore $\varphi(\mathcal{U}_0(A))$ is contained in $\mathcal{U}_0(B)$.

Conversely, if u belongs to $\mathcal{U}_0(B)$, then by Proposition 2.1.6,

$$u = \exp(ih_1) \cdot \exp(ih_2) \cdots \exp(ih_n)$$

for some self-adjoint elements h_j in B. As φ is surjective, there are elements x_j in A with $\varphi(x_j) = h_j$. Put $k_j = (x_j + x_j^*)/2$. Then $k_j = k_j^*$ and $\varphi(k_j) = h_j$. Set

$$v = \exp(ik_1) \cdot \exp(ik_2) \cdots \exp(ik_n).$$

Then v belongs to $\mathcal{U}_0(A)$ by Proposition 2.1.6, and $\varphi(v) = u$.

Part (ii) follows immediately from (i) and Lemma 2.1.5.

(iii). If $u \sim_h \varphi(v)$, then $u\varphi(v^*)$ belongs to $\mathcal{U}_0(B)$, and hence $u\varphi(v^*) = \varphi(w)$ for some w in $\mathcal{U}_0(A)$ by (i). Now $u = \varphi(wv)$ as desired. □

Let A be a unital C^*-algebra. The group of invertible elements in A is denoted by $\mathrm{GL}(A)$, and the set of elements in $\mathrm{GL}(A)$ that are homotopic to 1_A is denoted by $\mathrm{GL}_0(A)$. The group $\mathcal{U}(A)$ of unitary elements in A is a subgroup of $\mathrm{GL}(A)$. Part (ii) of the following proposition says that $\mathcal{U}(A)$ is a *retract* of $\mathrm{GL}(A)$. (A subspace X_0 of a topological space X is called a retract of X if there is a continuous map $\tau \colon X \to X_0$ satisfying $x \sim_h \tau(x)$ in X for all x in X, and $\tau(x) = x$ for all x in X_0.)

For each element a in A set $|a| = (a^*a)^{1/2}$. The element $|a|$ is called the *absolute value* of a.

Proposition 2.1.8. *Let A be a unital C^*-algebra.*

(i) *If z is an invertible element in A, then so is $|z|$, and $\omega(z) = z|z|^{-1}$ belongs to $\mathcal{U}(A)$. Clearly, $z = \omega(z)|z|$.*

(ii) *The map $\omega \colon \mathrm{GL}(A) \to \mathcal{U}(A)$ defined in (i) is continuous, $\omega(u) = u$ for every u in $\mathcal{U}(A)$, and $\omega(z) \sim_h z$ in $\mathrm{GL}(A)$ for every z in $\mathrm{GL}(A)$.*

(iii) *If u, v are unitary elements in $\mathcal{U}(A)$, and if $u \sim_h v$ in $\mathrm{GL}(A)$, then $u \sim_h v$ in $\mathcal{U}(A)$.*

Proof. (i). If z is invertible, then so are z^* and z^*z. It follows that $|z| = (z^*z)^{1/2}$ is invertible (with inverse $((z^*z)^{-1})^{1/2}$). Put $u = z|z|^{-1}$ $(= \omega(z))$. Then $z = u|z|$. Moreover, u is invertible, being a product of two invertible elements, and $u^{-1} = u^*$ because

$$u^*u = |z|^{-1}z^*z|z|^{-1} = |z|^{-1}|z|^2|z|^{-1} = 1.$$

(ii). Multiplication in a C^*-algebra is continuous, and so is the map $z \mapsto z^{-1}$ for z in $\mathrm{GL}(A)$. To show that ω is continuous, it therefore suffices to show that the map $a \mapsto |a|$ is continuous. This map is composed of the map

$a \mapsto a^*a$ and the map $h \mapsto h^{1/2}$ for h in A^+. The former of these maps is continuous because the involution and multiplication are continuous. It now suffices to show that $h \mapsto h^{1/2}$ is continuous on any bounded subset Ω of A^+. But this follows from Lemma 1.2.5 because each bounded subset Ω of A is contained in Ω_K, where $K = [0, R]$ and $R = \sup\{\|h\| : h \in \Omega\}$.

If u is a unitary element in A, then $|u| = 1$, and so $\omega(u) = u$.

Let z in $GL(A)$ be given, and put $z_t = \omega(z)(t|z| + (1-t)\cdot 1_A)$ for t in $[0,1]$. Then $\omega(z) = z_0$ and $z = z_1$. Since $|z|$ is positive and invertible, there is λ in $(0,1]$ for which $|z| \geqslant \lambda\cdot 1_A$. For each t in $[0,1]$ we have $t|z| + (1-t)\cdot 1_A \geqslant \lambda\cdot 1_A$. Thus $t|z| + (1-t)\cdot 1_A$ and z_t are invertible. The map $t \mapsto z_t$ is continuous, and so $\omega(z) = z_0 \sim_h z_1 = z$ in $GL(A)$.

(iii). If $t \mapsto z_t$ is a continuous path in $GL(A)$ from u to v, then $t \mapsto \omega(z_t)$ is a continuous path in $\mathcal{U}(A)$ from u to v. □

2.1.9 The polar decomposition. The factorization $z = \omega(z)|z|$ from Proposition 2.1.8 of an invertible element z is called the *(unitary) polar decomposition* for z. We shall often write u instead of $\omega(z)$, and the point of emphasis is then that each invertible element z in a unital C^*-algebra A can be written as $z = u|z|$ for a (unique) unitary element u in A.

2.1.10 The Carl Neumann series. Let A be a unital C^*-algebra. Then each a in A with $\|1_A - a\| < 1$ is invertible, and its inverse is given by the norm convergent series

$$(2.1) \qquad a^{-1} = 1_A + (1_A - a) + (1_A - a)^2 + (1_A - a)^3 + \cdots.$$

It follows in particular that the norm of a^{-1} can be estimated by

$$(2.2) \qquad \begin{aligned} \|a^{-1}\| &\leqslant 1 + \|1_A - a\| + \|1_A - a\|^2 + \|1_A - a\|^3 + \cdots \\ &= (1 - \|1_A - a\|)^{-1}. \end{aligned}$$

Proposition 2.1.11. *Let A be a unital C^*-algebra. Let a be an invertible element in A, and take b in A with $\|a - b\| < \|a^{-1}\|^{-1}$. Then b is invertible,*

$$\|b^{-1}\|^{-1} \geqslant \|a^{-1}\|^{-1} - \|a - b\|,$$

and $a \sim_h b$ in $GL(A)$.

Proof. From

$$\|1_A - a^{-1}b\| = \|a^{-1}(a - b)\| \leqslant \|a^{-1}\| \cdot \|a - b\| < 1,$$

we see that $a^{-1}b$ is invertible, and by (2.2),

$$\|(a^{-1}b)^{-1}\| \leqslant (1 - \|1_A - a^{-1}b\|)^{-1}.$$

This implies that b is invertible with inverse $b^{-1} = (a^{-1}b)^{-1}a^{-1}$, and

$$\|b^{-1}\|^{-1} \geqslant \|(a^{-1}b)^{-1}\|^{-1} \cdot \|a^{-1}\|^{-1} \geqslant (1 - \|1_A - a^{-1}b\|) \cdot \|a^{-1}\|^{-1}$$
$$\geqslant \|a^{-1}\|^{-1} - \|a - b\|.$$

To prove the last claim, put $c_t = (1-t)a + tb$ for t in $[0,1]$. Then $\|a - c_t\| = t\|a - b\| < \|a^{-1}\|^{-1}$, and therefore c_t is invertible by the first part of the proof. This shows that $a = c_0 \sim_h c_1 = b$ in $\mathrm{GL}(A)$. $\qquad\square$

2.2 Equivalence of projections

Definition 2.2.1. Recall that an element p in a C^*-algebra is a projection if $p = p^2 = p^*$. The set of all projections in a C^*-algebra A is denoted by $\mathcal{P}(A)$. From Paragraph 2.1.1 we have the homotopy equivalence relation $\sim_h$ on $\mathcal{P}(A)$. Consider also the following two equivalence relations on $\mathcal{P}(A)$:

- $p \sim q$ if there exists v in A with $p = v^*v$ and $q = vv^*$ (*Murray–von Neumann equivalence*),

- $p \sim_u q$ if there exists a unitary element u in $\mathcal{U}(\widetilde{A})$ with $q = upu^*$ (*unitary equivalence*).

An element v in A for which v^*v is a projection is called a *partial isometry*. As shown in Exercise 2.5, if v is a partial isometry, then vv^* is also a projection. The projection v^*v is called the *support projection* of v, and vv^* is called the *range projection* of v. Put $p = v^*v$ and $q = vv^*$. Then we have the following much used identities:

(2.3) $$v = qv = vp = qvp;$$

see Exercise 2.5.

We can use this to show that the Murray–von Neumann relation is transitive. Let p, q, and r be equivalent projections in A, and take partial isometries v and w such that $p = v^*v$, $q = vv^* = w^*w$, and $r = ww^*$. Put $z = wv$. Then

$$z^*z = v^*w^*wv = v^*qv = v^*v = p, \qquad zz^* = wvv^*w^* = wqw^* = ww^* = r,$$

so that $p \sim r$.

Proposition 2.2.2. *Let p, q be projections in a unital C^*-algebra A. The following conditions are equivalent:*

(i) $p \sim_u q$,

(ii) $q = upu^*$ *for some unitary u in A,*

(iii) $p \sim q$ *and* $1_A - p \sim 1_A - q$.

Proof. Let $1_{\widetilde{A}}$ denote the unit of $\widetilde{A}$, and put $f = 1_{\widetilde{A}} - 1_A$. Then $\widetilde{A} = A + \mathbb{C}f$, and $fa = af = 0$ for all a in A.

(i) $\Rightarrow$ (ii). Suppose that $q = zpz^*$ for some z in $\mathcal{U}(\widetilde{A})$. We have $z = u + \alpha f$ for some u in A and some α in $\mathbb{C}$. A straightforward calculation shows that u is unitary and that $q = upu^*$.

(ii) $\Rightarrow$ (iii). Suppose that $q = upu^*$ for some unitary u in $\mathcal{U}(A)$. Put $v = up$ and $w = u(1_A - p)$. Then

$$(2.4) \qquad v^*v = p, \quad vv^* = q, \quad w^*w = 1_A - p, \quad ww^* = 1_A - q.$$

(iii) $\Rightarrow$ (i). Suppose that there are partial isometries v and w satisfying (2.4) above. A direct calculation (using equation (2.3)) or Exercise 2.6 shows that $z = v + w + f$ is a unitary element in $\widetilde{A}$. By equation (2.3) we find that $zpz^* = vpv^* = vv^* = q$. $\square$

Lemma 2.2.3. *Let p be a projection in a C^*-algebra A, and let a be a self-adjoint element in A. Put $\delta = \|p - a\|$. Then*

$$\mathrm{sp}(a) \subseteq [-\delta, \delta] \cup [1 - \delta, 1 + \delta].$$

Proof. Since a is self-adjoint its spectrum consists of real numbers. Recall that $\mathrm{sp}(p)$ is contained in $\{0, 1\}$ (see Exercise 1.4). It suffices to show that if t is a real number whose distance, d, to the set $\{0, 1\}$ is strictly greater than δ, then t does not belong to $\mathrm{sp}(a)$. For such a real number t the element $p - t \cdot 1$ is invertible in $\widetilde{A}$ and

$$\|(p - t \cdot 1)^{-1}\| = \max\{|-t|^{-1}, |1 - t|^{-1}\} = d^{-1}.$$

It follows that

$$\|(p - t \cdot 1)^{-1}(a - t \cdot 1) - 1\| = \|(p - t \cdot 1)^{-1}(a - p)\| \leqslant d^{-1}\delta < 1.$$

This shows that $(p - t \cdot 1)^{-1}(a - t \cdot 1)$ is invertible. Hence $a - t \cdot 1$ is invertible, and t does not belong to $\mathrm{sp}(a)$. $\square$

Proposition 2.2.4. *If p, q are projections in a C^*-algebra A and $\|p-q\| < 1$, then $p \sim_h q$.*

Proof. Put $a_t = (1 - t)p + tq$ for t in $[0, 1]$. Then a_t is self-adjoint,

$$\min\{\|a_t - p\|, \|a_t - q\|\} \leqslant \|p - q\|/2 < 1/2,$$

and $t \mapsto a_t$ is continuous. Moreover, in the notation of Lemma 1.2.5, each a_t belongs to Ω_K by Lemma 2.2.3, where $K = [-\delta, \delta] \cup [1 - \delta, 1 + \delta]$ and $\delta = \|p - q\|/2 < 1/2$.

Define $f : K \to \mathbb{C}$ to be the continuous function which is zero on the interval $[-\delta, \delta]$ and 1 on the interval $[1 - \delta, 1 + \delta]$. (Note that these two intervals are disjoint.) Then $f(a_t)$ is a projection for each t in $[0, 1]$ because $f = f^2 = \bar{f}$. The path $t \mapsto f(a_t)$ is continuous by Lemma 1.2.5, and hence

$$p = f(p) = f(a_0) \sim_h f(a_1) = f(q) = q \quad \text{in} \quad \mathcal{P}(A).$$

$\square$

The next proposition says that if two self-adjoint elements in a unital C^*-algebra are similar, then they are unitarily equivalent.

Proposition 2.2.5. *Let a, b be self-adjoint elements in a unital C^*-algebra A, and suppose that $b = zaz^{-1}$ for some invertible element z in A. Let $z = u|z|$ be the polar decomposition for z with u in $\mathcal{U}(A)$, see Paragraph 2.1.9. Then $b = uau^*$.*

Proof. The equation $b = zaz^{-1}$ implies that $bz = za$ and, because a and b are self-adjoint, also $z^*b = az^*$. Hence

$$|z|^2 a = (z^*z)a = z^*bz = a(z^*z) = a|z|^2,$$

and so a commutes with $|z|^2$. Consequently a commutes with all elements in $C^*(1, |z|^2)$, and in particular a commutes with $|z|^{-1}$. It follows that

$$uau^* = z|z|^{-1}au^* = za|z|^{-1}u^* = bz|z|^{-1}u^* = buu^* = b,$$

as desired. $\square$

Proposition 2.2.6. *Let A be a C^*-algebra, and let p, q be projections in $\mathcal{P}(A)$. Then $p \sim_h q$ in $\mathcal{P}(A)$ if and only if there exists a unitary element u in $\mathcal{U}_0(\widetilde{A})$ such that $q = upu^*$.*

Proof. Throughout the proof, 1 denotes the unit of $\widetilde{A}$.

Suppose first that $q = upu^*$ for some u in $\mathcal{U}_0(\widetilde{A})$. Let $t \mapsto u_t$ be a continuous path of unitaries in $\widetilde{A}$ from 1 to u. Then, because A is an ideal in $\widetilde{A}$, $t \mapsto u_t p u_t^*$ is a continuous path of projections in A from p to q.

Conversely, if $p \sim_h q$, then there are projections $p = p_0, p_1, p_2, \ldots, p_n = q$ in A such that $\|p_{j+1} - p_j\| < 1/2$ for each j. It therefore suffices to prove the implication in the case where $\|p - q\| < 1/2$.

Put $z = pq + (1-p)(1-q)$. Then z belongs to $\widetilde{A}$, $pz = pq = zq$, and

$$\|z - 1\| = \|p(q - p) + (1 - p)((1 - q) - (1 - p))\| \leqslant 2\|q - p\| < 1.$$

Hence z is invertible and $z \sim_h 1$ in $GL(\widetilde{A})$ by Proposition 2.1.11. Let $z = u|z|$ be the unitary polar decomposition for z, see Paragraph 2.1.9. Then $p = uqu^*$ by Proposition 2.2.5. From the properties of the polar decomposition (Proposition 2.1.8) we have $u \sim_h z \sim_h 1$ in $GL(\widetilde{A})$, and this entails that u belongs to $\mathcal{U}_0(\widetilde{A})$ by Proposition 2.1.8 (iii). $\qquad\square$

We have three equivalence relations $\sim_h$, $\sim_u$, and $\sim$ on the set of projections in a C^*-algebra. It will be shown in Example 2.2.9 that these three equivalence relations are different from each other. Proposition 2.2.7 below says that homotopy is stronger than unitary equivalence which again is stronger than Murray–von Neumann equivalence; Proposition 2.2.8 says that the three relations actually are equal modulo passing to matrix algebras.

Proposition 2.2.7. *Let p, q be projections in a C^*-algebra A.*

(i) *If $p \sim_h q$, then $p \sim_u q$.*

(ii) *If $p \sim_u q$, then $p \sim q$.*

Proof. Part (i) is an immediate consequence of Proposition 2.2.6.

If $upu^* = q$ for some unitary u in $\widetilde{A}$, then $v = up$ belongs to A, $v^*v = p$, and $vv^* = q$. This proves (ii). $\qquad\square$

Proposition 2.2.8. *Let p, q be projections in a C^*-algebra A.*

(i) *If $p \sim q$, then $\begin{pmatrix} p & 0 \\ 0 & 0 \end{pmatrix} \sim_u \begin{pmatrix} q & 0 \\ 0 & 0 \end{pmatrix}$ in $M_2(A)$.*

(ii) *If $p \sim_u q$, then $\begin{pmatrix} p & 0 \\ 0 & 0 \end{pmatrix} \sim_h \begin{pmatrix} q & 0 \\ 0 & 0 \end{pmatrix}$ in $M_2(A)$.*

Proof. (i). Find v in A such that $p = v^*v$ and $q = vv^*$. Use (2.3) to see that

$$u = \begin{pmatrix} v & 1-q \\ 1-p & v^* \end{pmatrix}, \qquad w = \begin{pmatrix} q & 1-q \\ 1-q & q \end{pmatrix}$$

are unitary elements in $M_2(\widetilde{A})$. Since

$$wu \begin{pmatrix} p & 0 \\ 0 & 0 \end{pmatrix} u^*w^* = w \begin{pmatrix} q & 0 \\ 0 & 0 \end{pmatrix} w^* = \begin{pmatrix} q & 0 \\ 0 & 0 \end{pmatrix},$$

and since

$$wu = \begin{pmatrix} v + (1-q)(1-p) & (1-q)v^* \\ q(1-p) & (1-q) + qv^* \end{pmatrix}$$

belongs to $M_2(A)^\sim$, claim (i) is proved. (Notice that we must verify that the unitary element wu, which a priori belongs to $M_2(\widetilde{A})$, in fact belongs to the sub-C^*-algebra $M_2(A)^\sim$.)

(ii). Find u in $\mathcal{U}(\widetilde{A})$ such that $q = upu^*$. Lemma 2.1.5 provides us with a continuous path $t \mapsto w_t$ in $\mathcal{U}(M_2(\widetilde{A}))$ such that

$$w_0 = \begin{pmatrix} 1 & 0 \\ 0 & 1 \end{pmatrix}, \qquad w_1 = \begin{pmatrix} u & 0 \\ 0 & u^* \end{pmatrix}.$$

Put $e_t = w_t \operatorname{diag}(p,0) w_t^*$. Then each e_t belongs to $\mathcal{P}(M_2(A))$, the map $t \mapsto e_t$ is continuous, $e_0 = \operatorname{diag}(p,0)$, and $e_1 = \operatorname{diag}(q,0)$. $\qquad\square$

Example 2.2.9. Neither of the implications "$p \sim q \Rightarrow p \sim_u q$" and "$p \sim_u q \Rightarrow p \sim_h q$" holds in general.

An *isometry* in a unital C^*-algebra A is an element s such that $s^*s = 1$. Non-unitary isometries exist. Take for example the *unilateral shift* S on the Hilbert space $\ell^2(\mathbb{N})$ given by

$$(S\xi)(n) = \begin{cases} 0, & n = 1, \\ \xi(n-1), & n \geqslant 2, \end{cases}$$

for all ξ in $\ell^2(\mathbb{N})$. The adjoint of S is given by $(S^*\xi)(n) = \xi(n+1)$. It follows that $S^*S = I$ and hence that $\|S\xi\| = \|\xi\|$ for all ξ in $\ell^2(\mathbb{N})$, i.e., S is an isometry. Clearly, S is not surjective and therefore not unitary.

Let A be a unital C^*-algebra containing a non-unitary isometry s. By definition, $s^*s \sim ss^*$. The zero projection is not Murray–von Neumann equivalent to a non-zero projection, because if $v^*v = 0$, then $v = 0$ and so $vv^* = 0$.

Hence $1 - s^*s \ (= 0)$ is *not* equivalent to $1 - ss^* \ (\neq 0)$. By Proposition 2.2.2 we conclude that the projections s^*s and ss^* are *not* unitarily equivalent.

An example of a (unital) C^*-algebra A that contains projections p, q such that $p \sim_u q$ and $p \not\sim_h q$ is more complicated to construct. A unital C^*-algebra B, such that $M_2(B)$ contains a unitary element u which is not homotopic to $\mathrm{diag}(v, 1)$ for any unitary element v in $\mathcal{U}(B)$, will be found in Example 11.3.4. Put $A = M_2(B)$. The projections

$$p = \begin{pmatrix} 1 & 0 \\ 0 & 0 \end{pmatrix}, \qquad q = u \begin{pmatrix} 1 & 0 \\ 0 & 0 \end{pmatrix} u^*,$$

are unitarily equivalent in A, and they are not homotopic. Indeed, assume, to reach a contradiction, that $p \sim_h q$. Then, by Proposition 2.2.6, there is a unitary w in $\mathcal{U}_0(A)$ such that $wqw^* = p$, and hence $(wu)p = p(wu)$. As in Exercise 1.12 it follows that

$$wu = \begin{pmatrix} a & 0 \\ 0 & b \end{pmatrix}$$

for some elements a, b in $\mathcal{U}(B)$. Using the Whitehead lemma (Lemma 2.1.5) this leads to the contradiction

$$u \sim_h wu = \begin{pmatrix} a & 0 \\ 0 & b \end{pmatrix} \sim_h \begin{pmatrix} ab & 0 \\ 0 & 1 \end{pmatrix}.$$

2.2.10* Liftings. Suppose that A and B are C^*-algebras and that $\varphi \colon A \to B$ is a *surjective* *-homomorphism. Given an element b in B, an element a in A is called a *lift* of b if $\varphi(a) = b$. The set of all lifts of b is then the coset $a + \mathrm{Ker}(\varphi)$. We shall here be concerned with the possibility of lifting an element in B with a certain property to an element in A with the same property. Along this line we have the following results.

(i) Every element b in B has a lift to an element a in A with $\|a\| = \|b\|$.

(ii) Every self-adjoint element b in B lifts to a self-adjoint element a in A. Moreover, the self-adjoint lift a can be chosen such that $\|a\| = \|b\|$.

(iii) Every positive element b in B lifts to a positive element a in A. Moreover, the positive lift a can be chosen such that $\|a\| = \|b\|$.

(iv) A normal element in B does not in general lift to a normal element in A.

(v) A projection in B does not in general lift to a projection in A.

(vi) A unitary element in B does not in general lift to a unitary element in A, when A and B are unital C^*-algebras. (See Exercise 2.12, and also Lemma 2.1.7, Exercises 8.6 and 9.2.)

Proof. (ii). Let x in A be any lift of b, and set $a_0 = (x + x^*)/2$. Then a_0 is self-adjoint and $\varphi(a_0) = b$. To arrange that the lift has the same norm as b, let $f \colon \mathbb{R} \to \mathbb{R}$ be the continuous function given by

$$f(t) = \begin{cases} -\|b\|, & t \leqslant -\|b\|, \\ t, & -\|b\| \leqslant t \leqslant \|b\|, \\ \|b\|, & t \geqslant \|b\|, \end{cases}$$

and put $a = f(a_0)$. Then a is normal, being a continuous function of a normal element, and

$$\mathrm{sp}(a) = \{f(t) : t \in \mathrm{sp}(a_0)\} \subseteq [-\|b\|, \|b\|].$$

This shows that a is self-adjoint, and $\|a\| \leqslant \|b\|$ because the norm of a is its spectral radius. Also,

$$\varphi(a) = \varphi(f(a_0)) = f(\varphi(a_0)) = f(b) = b,$$

because $f(t) = t$ for each t in $\mathrm{sp}(b)$. As φ is a *-homomorphism, $\|\varphi\| \leqslant 1$, and we conclude that $\|a\| = \|b\|$.

(i). Let b be an element in B, and put

$$y = \begin{pmatrix} 0 & b \\ b^* & 0 \end{pmatrix}.$$

Then y is a self-adjoint element in $M_2(B)$, and

$$\|y\|^2 = \|y^*y\| = \left\| \begin{pmatrix} bb^* & 0 \\ 0 & b^*b \end{pmatrix} \right\| = \max\{\|bb^*\|, \|b^*b\|\} = \|b\|^2.$$

Consult Exercise 1.15 regarding the third equality sign. It follows from (ii) that there is a self-adjoint lift

$$x = \begin{pmatrix} x_{11} & x_{12} \\ x_{21} & x_{22} \end{pmatrix} \in M_2(A)$$

of y with $\|x\| = \|y\|$. The element $a = x_{12}$ in A is a lift of b, and by (1.4)

$\|a\| \leqslant \|x\| = \|y\| = \|b\|$. As in the proof of (ii), necessarily $\|b\| \leqslant \|a\|$, and so $\|a\| = \|b\|$.

(iii). Let x in A be any lift of b, and set $a_0 = (x^*x)^{1/2}$. Then a_0 is positive and
$$\varphi(a_0) = (\varphi(x)^*\varphi(x))^{1/2} = (b^*b)^{1/2} = b.$$

As in the proof of (ii), put $a = f(a_0)$. Then a is normal, $\varphi(a) = b$, and $\mathrm{sp}(a) \subseteq [0, \|b\|]$. Hence a is positive and $\|a\| = \|b\|$.

(iv). See Exercise 9.4 (iii).

(v). Let $A = C([0,1])$, let $B = \mathbb{C} \oplus \mathbb{C}$, let $\varphi(f) = (f(0), f(1))$, and let $q = (0,1)$. Then q is a projection in B, and there is no projection p in A such that $\varphi(p) = q$.

(vi). See Exercise 2.12 (ii). $\square$

2.3 Semigroups of projections

Definition 2.3.1 (The semigroup $\mathcal{P}_\infty(A)$). Put

$$\mathcal{P}_n(A) = \mathcal{P}(M_n(A)), \qquad \mathcal{P}_\infty(A) = \bigcup_{n=1}^{\infty} \mathcal{P}_n(A),$$

where A is a C^*-algebra and n is a positive integer. We view the sets $\mathcal{P}_n(A)$, $n \in \mathbb{N}$, as being pairwise disjoint.

Define the relation $\sim_0$ on $\mathcal{P}_\infty(A)$ as follows. Suppose that p is a projection in $\mathcal{P}_n(A)$ and q is a projection in $\mathcal{P}_m(A)$. Then $p \sim_0 q$ if there is an element v in $M_{m,n}(A)$ with $p = v^*v$ and $q = vv^*$. Here, $M_{m,n}(A)$ is the set of all rectangular $m \times n$ matrices

$$\begin{pmatrix} a_{11} & a_{12} & \cdots & a_{1n} \\ a_{21} & a_{22} & \cdots & a_{2n} \\ \vdots & \vdots & \ddots & \vdots \\ a_{m1} & a_{m2} & \cdots & a_{mn} \end{pmatrix}$$

with entries a_{ij} in A. The adjoint v^* of an element v in $M_{m,n}(A)$ is the element in $M_{n,m}(A)$ obtained by transposing and taking adjoints entry-wise. The products vv^* and v^*v are the usual matrix products.

Define a binary operation $\oplus$ on $\mathcal{P}_\infty(A)$ by

$$p \oplus q = \text{diag}(p, q) = \begin{pmatrix} p & 0 \\ 0 & q \end{pmatrix},$$

so that $p \oplus q$ belongs to $\mathcal{P}_{n+m}(A)$ when p is in $\mathcal{P}_n(A)$ and q is in $\mathcal{P}_m(A)$.

The relation $\sim_0$ is an equivalence relation on $\mathcal{P}_\infty(A)$. It combines the Murray–von Neumann equivalence relation $\sim$ with an identification of projections in different sized matrix algebras over A. If p and q both belong to $\mathcal{P}_n(A)$ for some n, then $p \sim_0 q$ if and only if p and q are Murray–von Neumann equivalent.

The relation $\sim_0$ is named after the K_0-group it will help defining.

Proposition 2.3.2. *Let p, q, r, p', q' be projections in $\mathcal{P}_\infty(A)$ for some C^*-algebra A.*

(i) $p \sim_0 p \oplus 0_n$ *for every natural number n, where 0_n is the zero element of $M_n(A)$,*

(ii) *if $p \sim_0 p'$ and $q \sim_0 q'$, then $p \oplus q \sim_0 p' \oplus q'$,*

(iii) $p \oplus q \sim_0 q \oplus p$,

(iv) *if p, q are projections in $\mathcal{P}_n(A)$ such that $pq = 0$, then $p + q$ is a projection and $p + q \sim_0 p \oplus q$,*

(v) $(p \oplus q) \oplus r = p \oplus (q \oplus r)$.

Proof. (i). Let m, n be positive integers, and let p be a projection in $\mathcal{P}_m(A)$. Put

$$u_1 = \begin{pmatrix} p \\ 0 \end{pmatrix} \in M_{m+n,m}(A).$$

Then $p = u_1^* u_1 \sim_0 u_1 u_1^* = p \oplus 0_n$.

(ii). If $p \sim_0 p'$ and $q \sim_0 q'$, then there exist v, w such that

$$p = v^* v, \quad p' = vv^*, \quad q = w^* w, \quad q' = ww^*.$$

Put $u_2 = \text{diag}(v, w)$. Then $p \oplus q = u_2^* u_2 \sim_0 u_2 u_2^* = p' \oplus q'$.

(iii). Suppose that p belongs to $\mathcal{P}_n(A)$ and q belongs to $\mathcal{P}_m(A)$, and put

$$u_3 = \begin{pmatrix} 0_{n,m} & q \\ p & 0_{m,n} \end{pmatrix}$$

where $0_{k,l}$ is the zero element in $M_{k,l}(A)$. Then u_3 belongs to $M_{n+m}(A)$, and $p \oplus q = u_3^* u_3 \sim u_3 u_3^* = q \oplus p$.

(iv). If $n = m$ and $pq = 0$, then $p + q$ is a projection (see Exercise 2.4). Put

$$u_4 = \begin{pmatrix} p \\ q \end{pmatrix} \in M_{2n,n}(A).$$

Then $p + q = u_4^* u_4 \sim_0 u_4 u_4^* = p \oplus q$.

Part (v) is trivial.   □

Definition 2.3.3 (The semigroup $\mathcal{D}(A)$). With $(\mathcal{P}_\infty(A), \sim_0, \oplus)$ as in Definition 2.3.1, set

$$\mathcal{D}(A) = \mathcal{P}_\infty(A)/\sim_0 .$$

For each p in $\mathcal{P}_\infty(A)$ let $[p]_\mathcal{D}$ in $\mathcal{D}(A)$ denote the equivalence class containing p. Define addition on $\mathcal{D}(A)$ by

$$[p]_\mathcal{D} + [q]_\mathcal{D} = [p \oplus q]_\mathcal{D}, \qquad p, q \in \mathcal{P}_\infty(A).$$

It follows from Proposition 2.3.2 that this operation is well-defined and that $(\mathcal{D}(A), +)$ is an Abelian semigroup.

In the next chapter we shall, for each unital C^*-algebra A, construct an Abelian group $K_0(A)$ from the semigroup $(\mathcal{D}(A), +)$.

2.4   Exercises

Exercise 2.1. Show that $\|p - q\| \leqslant 1$ for every pair of projections p, q in a C^*-algebra, and show that $\|u - v\| \leqslant 2$ for every pair of unitary elements u, v in a unital C^*-algebra.

Exercise 2.2. Let a be a self-adjoint element in a unital C^*-algebra A with $\|a\| \leqslant 1$. Show that

$$a + i\sqrt{1 - a^2}, \qquad a - i\sqrt{1 - a^2}$$

are unitary elements in A. Conclude from this and Exercise 1.7 that each element in a unital C^*-algebra is a linear combination of four unitaries.

Is it true that each element in a C^*-algebra can be written as a linear combination of projections in the C^*-algebra?

Exercise 2.3. Let A be a unital C^*-algebra, and consider the upper triangular matrix

$$a = \begin{pmatrix} 1 & a_{12} & a_{13} & \cdots & a_{1n} \\ 0 & 1 & a_{23} & \cdots & a_{2n} \\ 0 & 0 & 1 & \cdots & a_{3n} \\ \vdots & \vdots & \vdots & \ddots & \vdots \\ 0 & 0 & 0 & \cdots & 1 \end{pmatrix}$$

in $M_n(A)$. Use Exercise 1.14 to see that a is invertible. Show that $a \sim_h 1$ in $GL(M_n(A))$.

Exercise 2.4. Two projections p, q in a C^*-algebra A are said to be *orthogonal to each other*, or mutually orthogonal (in symbols $p \perp q$), if $pq = 0$. Show that the following three conditions are equivalent:

(i) $p \perp q$,

(ii) $p + q$ is a projection,

(iii) $p + q \leqslant 1$.

[Hint: To see that (iii) implies (i), observe that (iii) implies $p(p + q)p \leqslant p$.] A family $p_1, p_2, \ldots, p_n$ of projections in A is said to be mutually orthogonal if $p_i \perp p_j$ whenever $i \neq j$. Use the first part of this exercise to show that the following conditions are equivalent:

(i) $p_1, p_2, \ldots, p_n$ are mutually orthogonal,

(ii) $p_1 + p_2 + \cdots + p_n$ is a projection,

(iii) $p_1 + p_2 + \cdots + p_n \leqslant 1$.

Exercise 2.5. Let A be a C^*-algebra, and let v in A be such that v^*v is a projection (i.e., v is a partial isometry). Show that $v = vv^*v$, and conclude that vv^* is a projection.
  Put $p = v^*v$ and $q = vv^*$. Show that $v = qv = vp = qvp$.
  [Hint: To show that $v = vv^*v$ put $z = (1 - vv^*)v$ and calculate z^*z.]

Exercise 2.6. Let $v_1, v_2, \ldots, v_n$ be partial isometries in a unital C^*-algebra A, and suppose that

$$\sum_{j=1}^{n} v_j^* v_j = 1_A = \sum_{j=1}^{n} v_j v_j^*.$$

Show that $\sum_{j=1}^{n} v_j$ is unitary. [Hint: Use Exercise 2.4 and equation (2.3).]

Exercise 2.7. Let $\varepsilon > 0$ be given. Show that there exists $\delta > 0$ with the following property. If A is a C^*-algebra and if a is an element in A such that $\|a - a^*\| \leqslant \delta$ and $\|a^2 - a\| \leqslant \delta$, then there is a projection p in A with $\|a - p\| \leqslant \varepsilon$. In other words, an element a in A, which is almost a projection, is close to a projection in A. [Hint: We need only consider the case where $\varepsilon < 1/2$. Put $b = (a + a^*)/2$. Show that the spectrum of b is contained in $[-\varepsilon, \varepsilon] \cup [1 - \varepsilon, 1 + \varepsilon]$ if $\|b - b^2\| \leqslant \varepsilon - \varepsilon^2$, and put $p = f(b)$ for a suitably chosen continuous function f.]

Show that for each $\varepsilon > 0$ there exists $\delta > 0$ with the following property. If A is a C^*-algebra, B is a sub-C^*-algebra of A, and p is a projection in A such that $\|p - b\| \leqslant \delta$ for some b in B, then there is a projection q in B with $\|p - q\| \leqslant \varepsilon$.

Exercise 2.8. Let $\varepsilon > 0$ be given. Show that there exists $\delta > 0$ with the following property. If A is a unital C^*-algebra and a is an element in A such that $\|aa^* - 1_A\| \leqslant \delta$ and $\|a^*a - 1_A\| \leqslant \delta$, then there is a unitary element u in A with $\|a - u\| \leqslant \varepsilon$. In other words, an element a in A, which is almost unitary, is close to a unitary element in A. [Hint: Look at Proposition 2.1.8 and Paragraph 2.1.9.]

Show that for each $\varepsilon > 0$ there exists $\delta > 0$ with the following property. If A is a unital C^*-algebra, B is a sub-C^*-algebra of A containing the unit of A, and u is a unitary element in A such that $\|u - b\| \leqslant \delta$ for some element b in B, then there is a unitary element v in B with $\|u - v\| \leqslant \varepsilon$.

Exercise 2.9. Let $\mathrm{Tr}\colon M_n(\mathbb{C}) \to \mathbb{C}$ be the standard trace given by

$$\mathrm{Tr}\begin{pmatrix} \alpha_{11} & \alpha_{12} & \cdots & \alpha_{1n} \\ \alpha_{21} & \alpha_{22} & \cdots & \alpha_{2n} \\ \vdots & \vdots & \ddots & \vdots \\ \alpha_{n1} & \alpha_{n2} & \cdots & \alpha_{nn} \end{pmatrix} = \sum_{j=1}^{n} \alpha_{jj}.$$

Let p, q be projections in $M_n(\mathbb{C})$. Show that the following are equivalent:

(i) $p \sim q$,

(ii) $\mathrm{Tr}(p) = \mathrm{Tr}(q)$,

(iii) $\dim(p(\mathbb{C}^n)) = \dim(q(\mathbb{C}^n))$.

Use this to show that

$$\mathcal{D}(\mathbb{C}) \cong \{0, 1, 2, \ldots\} = \mathbb{Z}^+,$$

when $\mathbb{Z}^+$ is equipped with the usual addition.

Show finally that the implications

$$\text{``}p \sim q \Rightarrow p \sim_u q\text{''} \quad \text{and} \quad \text{``}p \sim q \Rightarrow p \sim_h q\text{''}$$

hold for all projections p, q in $M_n(\mathbb{C})$.

Exercise 2.10. Let p, q be projections in $B(H)$, where H is an infinite dimensional separable Hilbert space.

(i) Show that $p \sim q$ if and only if $\dim(p(H)) = \dim(q(H))$.

(ii) Show that $p \sim_u q$ if and only if

$$\dim(p(H)) = \dim(q(H)) \quad \text{and} \quad \dim(p(H)^\perp) = \dim(q(H)^\perp).$$

This result is in Example 3.3.3 used to show that

$$\mathcal{D}(B(H)) \cong \{0, 1, 2, \ldots, \infty\} = \mathbb{Z}^+ \cup \{\infty\},$$

where addition on $\mathbb{Z}^+ \cup \{\infty\}$ is the usual addition on $\mathbb{Z}^+$ and where $\infty + n = n + \infty = \infty$ for all n in $\mathbb{Z}^+ \cup \{\infty\}$.

Exercise 2.11. Show that $\mathcal{D}(\mathbb{C} \oplus \mathbb{C})$ is isomorphic to the additive semigroup $\mathbb{Z}^+ \oplus \mathbb{Z}^+$.

Exercise 2.12. Consider the short exact sequence

$$0 \longrightarrow C_0(\mathbb{R}^2) \overset{\varphi}{\longrightarrow} C(\mathbb{D}) \overset{\psi}{\longrightarrow} C(\mathbb{T}) \longrightarrow 0,$$

where $\mathbb{D} = \{z \in \mathbb{C} : |z| \leqslant 1\}$, where ψ is the restriction mapping, and where φ is obtained by identifying $\mathbb{D} \setminus \mathbb{T}$ with $\mathbb{R}^2$. (You may replace $\mathbb{R}^2$ with $\mathbb{D} \setminus \mathbb{T}$ if you wish.) Let v in $C(\mathbb{T})$ be given by $v(z) = z$ for all z in $\mathbb{T}$.

(i) Show that v is unitary.

(ii) Show that v does not lift to a unitary in $C(\mathbb{D})$, i.e., there is no unitary u in $C(\mathbb{D})$ such that $\psi(u) = v$. [Hint: Use Brouwer's fixed point theorem which says that each continuous function $f : \mathbb{D} \to \mathbb{D}$ has a fixed point.]

(iii) Conclude that v does not belong to $\mathcal{U}_0(C(\mathbb{T}))$, and that there exist unitaries v_1, v_2 in $C(\mathbb{T})$ such that $v_1 \nsim_h v_2$. Show that there is no self-adjoint element h in $C(\mathbb{T})$ for which $v = \exp(ih)$.

Chapter 3

The K_0-Group of a Unital C^*-Algebra

An Abelian group $K_0(A)$ is associated to each unital C^*-algebra A. The group $K_0(A)$ arises from the Abelian semigroup $(\mathcal{D}(A), +)$ (defined in Chapter 2) and the Grothendieck construction (described below). We shall see that K_0 is a functor from the category of unital C^*-algebras to the category of Abelian groups, and some of the properties of K_0 will be derived. Some examples of K_0-groups can be found at the end of the chapter.

We extend K_0 to a functor from the category of all C^*-algebras (unital or not) in Chapter 4.

3.1 Definition of the K_0-group of a unital C^*-algebra

3.1.1 The Grothendieck construction. One can associate an Abelian group to every Abelian semigroup in a way analogous to how one obtains the integers from the natural numbers, and in much the same way as one obtains the rational numbers from the integers. We describe here how this works; the proofs of various statements along the way are deferred to the next Paragraph.

Let $(S, +)$ be an Abelian semigroup. Define an equivalence relation $\sim$ on $S \times S$ by $(x_1, y_1) \sim (x_2, y_2)$ if there exists z in S such that $x_1 + y_2 + z = x_2 + y_1 + z$. That $\sim$ is an equivalence relation is proved in Paragraph 3.1.2.

Write $G(S)$ for the quotient $(S \times S)/\sim$, and let $\langle x, y \rangle$ denote the equivalence class in $G(S)$ containing (x, y) in $S \times S$. The operation

$$\langle x_1, y_1 \rangle + \langle x_2, y_2 \rangle = \langle x_1 + x_2, y_1 + y_2 \rangle$$

is well-defined and turns $(G(S), +)$ into an Abelian group. Notice that $-\langle x, y \rangle = \langle y, x \rangle$ and that $\langle x, x \rangle = 0$ for all x, y in S. The group $G(S)$ is called the *Grothendieck group* of S.

Take y in S. The map

$$\gamma_S : S \to G(S), \quad x \mapsto \langle x + y, y \rangle,$$

is independent of the choice of y, and γ_S is additive. It is called the *Grothendieck map*.

The semigroup $(S, +)$ is said to have the *cancellation property* if, whenever x, y, and z are elements in S with $x + z = y + z$, it follows that $x = y$.

The Grothendieck construction has the following properties.

(i) *The universal property.* If H is an Abelian group, and if $\varphi \colon S \to H$ is an additive map, then there is one and only one group homomorphism $\psi \colon G(S) \to H$ making the diagram

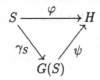

commutative.

(ii) *Functoriality.* To every additive map $\varphi \colon S \to T$ between semigroups S and T there is precisely one group homomorphism $G(\varphi) \colon G(S) \to G(T)$ making the diagram

$$
\begin{array}{ccc}
S & \overset{\varphi}{\longrightarrow} & T \\
\gamma_S \downarrow & & \downarrow \gamma_T \\
G(S) & \underset{G(\varphi)}{\longrightarrow} & G(T)
\end{array}
$$

commutative.

(iii) $G(S) = \{\gamma_S(x) - \gamma_S(y) : x, y \in S\}$.

(iv) Let x, y be elements in S. Then $\gamma_S(x) = \gamma_S(y)$ if and only if $x+z = y+z$ for some z in S.

(v) The Grothendieck map $\gamma_S \colon S \to G(S)$ is injective if and only if S has the cancellation property.

(vi) Let $(H, +)$ be an Abelian group, and let S be a non-empty subset of H. If S is closed under addition, then $(S, +)$ is an Abelian semigroup with the cancellation property, $G(S)$ is isomorphic to the subgroup H_0 generated by S, and $H_0 = \{x - y : x, y \in S\}$.

3.1.2 Some proofs. We shall here prove some of the claims from Paragraph 3.1.1.

The relation $\sim$ *is an equivalence relation on S*. It is clear that $\sim$ is reflexive and symmetric. To see that $\sim$ is transitive, suppose that $(x_1, y_1) \sim (x_2, y_2)$ and $(x_2, y_2) \sim (x_3, y_3)$. Then

$$x_1 + y_2 + z = x_2 + y_1 + z, \qquad x_2 + y_3 + w = x_3 + y_2 + w$$

for some z, w in S. It follows that

$$x_1 + y_3 + (y_2 + z + w) = x_2 + y_1 + z + y_3 + w = x_3 + y_1 + (y_2 + z + w),$$

and so $(x_1, y_1) \sim (x_3, y_3)$.

(iii). Each element in $G(S)$ has the form $\langle x, y \rangle$ for some x, y in S, and

$$\langle x, y \rangle = \langle x + y, y \rangle - \langle x + y, x \rangle = \gamma_S(x) - \gamma_S(y).$$

(iv). To see the "if" part, use that γ_S is additive and that $G(S)$ is a group. Suppose conversely that $\gamma_S(x) = \gamma_S(y)$. Then $\langle x + y, y \rangle = \langle y + x, x \rangle$, and therefore $(x + y) + x + w = (y + x) + y + w$ for some w in S. This shows that $x + z = y + z$, where $z = x + y + w$.

Part (v) is an immediate consequence of (iv).

(i). If $\langle x_1, y_1 \rangle = \langle x_2, y_2 \rangle$, then $x_1 + y_2 + z = x_2 + y_1 + z$ for some z in S, and hence

$$\varphi(x_1) + \varphi(y_2) + \varphi(z) = \varphi(x_2) + \varphi(y_1) + \varphi(z)$$

in H, which implies that $\varphi(x_1) - \varphi(y_1) = \varphi(x_2) - \varphi(y_2)$. Therefore the map $\psi \colon G(S) \to H$ given by $\psi(\langle x, y \rangle) = \varphi(x) - \varphi(y)$ is well-defined, and it makes the diagram in (i) commutative. Additivity of ψ follows easily from additivity of φ. The uniqueness of ψ follows from (iii).

(ii). By (i), the additive map $\gamma_T \circ \varphi \colon S \to G(T)$ factors uniquely through a group homomorphism $G(\varphi) \colon G(S) \to G(T)$.

(vi). A non-empty subset S of an Abelian group H which is closed under addition is an Abelian semigroup with the cancellation property. By (i) applied to the inclusion map $\iota \colon S \to H$, there is a group homomorphism $\psi \colon G(S) \to H$ such that $\psi(\gamma_S(x)) = x$ for all x in S. The image of ψ is precisely $\{x - y : x, y \in S\} = H_0$ by (iii). If $\psi(\gamma_S(x) - \gamma_S(y)) = 0$, then $x = y$ and so $\gamma_S(x) - \gamma_S(y) = 0$. Hence ψ is injective.

Example 3.1.3.

(i) The Grothendieck group of the Abelian semigroup $(\mathbb{Z}^+, +)$ is (isomorphic to) $(\mathbb{Z}, +)$ by Paragraph 3.1.1 (vi); and $(\mathbb{Z}^+, +)$ has the cancellation property. (See also Exercise 2.9.)

(ii) The Grothendieck group of the Abelian semigroup $(\mathbb{Z}^+ \cup \{\infty\}, +)$ is $\{0\}$, and $(\mathbb{Z}^+ \cup \{\infty\}, +)$ is a semigroup that does not have the cancellation property. (See also Exercise 2.10.)

Definition 3.1.4 (The K_0-group for a unital C^*-algebra). Let A be a unital C^*-algebra, and let $(\mathcal{D}(A), +)$ be the Abelian semigroup from Definition 2.3.3. Define $K_0(A)$ to be the Grothendieck group of $\mathcal{D}(A)$, i.e.,

$$K_0(A) = G(\mathcal{D}(A)).$$

Define $[\,\cdot\,]_0 \colon \mathcal{P}_\infty(A) \to K_0(A)$ by

(3.1) $[p]_0 = \gamma([p]_\mathcal{D}) \in K_0(A), \quad p \in \mathcal{P}_\infty(A),$

where $\gamma \colon \mathcal{D}(A) \to K_0(A)$ is the Grothendieck map.

3.1.5 The group $K_{00}(A)$. The definition of $K_0(A)$ given above also makes sense for non-unital C^*-algebras. With the notation from Blackadar's book, [3], let $K_{00}(A)$ denote the Grothendieck group of the semigroup $\mathcal{D}(A)$, i.e., $K_{00}(A) = G(\mathcal{D}(A))$ for every (unital or non-unital) C^*-algebra A.

Let $[\,\cdot\,]_{00} \colon \mathcal{P}_\infty(A) \to K_{00}(A)$ be given by $[p]_{00} = \gamma([p]_\mathcal{D})$ for each p in $\mathcal{P}_\infty(A)$. Then $K_0(A)$ and $K_{00}(A)$ are equal by definition for unital C^*-algebras A. However, the groups $K_0(A)$ and $K_{00}(A)$ are in general *not* equal for *non-unital* C^*-algebras A (with the definition of K_0 for non-unital C^*-algebras given in Chapter 4). The functor K_{00} has the serious weakness of not being half exact (Example 3.3.9).

Definition 3.1.6 (Stable equivalence). Define a relation $\sim_s$ on $\mathcal{P}_\infty(A)$ as follows. If p, q are projections in $\mathcal{P}_\infty(A)$, then $p \sim_s q$ if and only if $p \oplus r \sim_0 q \oplus r$ for some projection r in $\mathcal{P}_\infty(A)$. The relation $\sim_s$ is called *stable equivalence*.

Suppose that A is unital and that p, q are projections in $\mathcal{P}_\infty(A)$. Denote by 1_n the unit of $M_n(A)$. Then $p \sim_s q$ if and only if $p \oplus 1_n \sim_0 q \oplus 1_n$ for some positive integer n. Indeed, if $p \oplus r \sim_0 q \oplus r$ for some r in $\mathcal{P}_n(A)$, then

$$ p \oplus 1_n \sim_0 p \oplus r \oplus (1_n - r) \sim_0 q \oplus r \oplus (1_n - r) \sim_0 q \oplus 1_n. $$

The standard picture of K_0, described in the two propositions below, is a concrete and useful description of the K_0-group of a unital C^*-algebra. Proposition 3.1.8 shows that Proposition 3.1.7 (i), (ii), and (iii) form a *universal property* of K_0.

Proposition 3.1.7 (The standard picture of K_0 — the unital case). *Let A be a unital C^*-algebra. Then*

(3.2)
$$ \begin{aligned} K_0(A) &= \{[p]_0 - [q]_0 : p, q \in \mathcal{P}_\infty(A)\} \\ &= \{[p]_0 - [q]_0 : p, q \in \mathcal{P}_n(A), \ n \in \mathbb{N}\}. \end{aligned} $$

Moreover,

  (i) *$[p \oplus q]_0 = [p]_0 + [q]_0$ for all projections p, q in $\mathcal{P}_\infty(A)$,*

  (ii) *$[0_A]_0 = 0$, where 0_A is the zero projection in A. ,*

  (iii) *if p, q belong to $\mathcal{P}_n(A)$ for some n and $p \sim_h q$ in $\mathcal{P}_n(A)$, then $[p]_0 = [q]_0$,*

  (iv) *if p, q are mutually orthogonal projections in $\mathcal{P}_n(A)$, then $[p + q]_0 = [p]_0 + [q]_0$,*

  (v) *for all p, q in $\mathcal{P}_\infty(A)$, $[p]_0 = [q]_0$ if and only if $p \sim_s q$.*

Proof. The first identity in (3.2) follows from Paragraph 3.1.1 (iii). Hence, if g is an element in $K_0(A)$, then $g = [p']_0 - [q']_0$ for some p' in $\mathcal{P}_k(A)$ and q' in $\mathcal{P}_l(A)$. Choose n greater than k and l, and set $p = p' \oplus 0_{n-k}$ and $q = q' \oplus 0_{n-l}$. Then p, q are projections in $\mathcal{P}_n(A)$ with $p \sim_0 p'$ and $q \sim_0 q'$ by Proposition 2.3.2 (i). It follows that $g = [p]_0 - [q]_0$.

  (i). We have

$$ [p \oplus q]_0 = \gamma([p \oplus q]_\mathcal{D}) = \gamma([p]_\mathcal{D} + [q]_\mathcal{D}) = \gamma([p]_\mathcal{D}) + \gamma([q]_\mathcal{D}) = [p]_0 + [q]_0. $$

  (ii). Since $0_A \oplus 0_A \sim_0 0_A$, (i) yields that $[0_A]_0 + [0_A]_0 = [0_A]_0$, and hence $[0_A]_0 = 0$.

Part (iii) follows from the implications

$$p \sim_h q \;\Rightarrow\; p \sim q \;\Rightarrow\; p \sim_0 q \;\Longleftrightarrow\; [p]_\mathcal{D} = [q]_\mathcal{D} \;\Rightarrow\; [p]_0 = [q]_0,$$

where the first two relations are defined only when p and q are in the same matrix algebra over A, and the other three relations are defined for all p, q in $\mathcal{P}_\infty(A)$. The first implication holds by Proposition 2.2.7.

(iv). We have $p+q \sim_0 p \oplus q$ (Proposition 2.3.2), and so $[p+q]_0 = [p \oplus q]_0 = [p]_0 + [q]_0$ by (i).

(v). If $[p]_0 = [q]_0$, then there is a projection r in $\mathcal{P}_\infty(A)$ with $[p]_\mathcal{D} + [r]_\mathcal{D} = [q]_\mathcal{D} + [r]_\mathcal{D}$ by Paragraph 3.1.1 (iv). Hence $[p \oplus r]_\mathcal{D} = [q \oplus r]_\mathcal{D}$, so that $p \oplus r \sim_0 q \oplus r$, and therefore $p \sim_s q$.

Conversely, if $p \sim_s q$, then there is a projection r in $\mathcal{P}_\infty(A)$ with $p \oplus r \sim_0 q \oplus r$. By (i) we get $[p]_0 + [r]_0 = [q]_0 + [r]_0$, and because $K_0(A)$ is a group, we conclude that $[p]_0 = [q]_0$. $\square$

Proposition 3.1.8 (Universal property of K_0). *Let A be a unital C^*-algebra, let G be an Abelian group, and suppose that $\nu\colon \mathcal{P}_\infty(A) \to G$ is a map that satisfies*

(i) *$\nu(p \oplus q) = \nu(p) + \nu(q)$ for all projections p, q in $\mathcal{P}_\infty(A)$,*

(ii) *$\nu(0_A) = 0$,*

(iii) *if p, q belong to $\mathcal{P}_n(A)$ for some n and $p \sim_h q$ in $\mathcal{P}_n(A)$, then $\nu(p) = \nu(q)$.*

Then there is a unique group homomorphism $\alpha\colon K_0(A) \to G$ which makes the diagram

(3.3)

$$
\begin{array}{c}
\mathcal{P}_\infty(A) \\
{\scriptstyle [\,\cdot\,]_0} \downarrow \qquad \searrow {\scriptstyle \nu} \\
K_0(A) \xrightarrow{\ \ \alpha\ \ } G
\end{array}
$$

commutative.

One can replace property (iii) in Proposition 3.1.8 with a number of other properties, see Exercise 3.2.

Proof of proposition. We first show that if p, q are projections in $\mathcal{P}_\infty(A)$ and if $p \sim_0 q$, then $\nu(p) = \nu(q)$. Find k, l in $\mathbb{N}$ such that p belongs to $\mathcal{P}_k(A)$

and q belongs to $\mathcal{P}_l(A)$. Let n be an integer greater than both k and l, and put $p' = p \oplus 0_{n-k}$ and $q' = q \oplus 0_{n-l}$. Then p', q' both belong to $\mathcal{P}_n(A)$, $p' \sim_0 p \sim_0 q \sim_0 q'$, and so $p' \sim q'$. By Proposition 2.2.8, $p' \oplus 0_{3n} \sim_h q' \oplus 0_{3n}$ in $\mathcal{P}_{4n}(A)$. We conclude that

$$\nu(p) = \nu(p) + \underbrace{\nu(0) + \cdots + \nu(0)}_{4n-k} = \nu(p' \oplus 0_{3n}) = \nu(q' \oplus 0_{3n}) = \nu(q).$$

It follows that the map $\beta \colon \mathcal{D}(A) \to G$ given by $\beta([p]_\mathcal{D}) = \nu(p)$ is well-defined. Additivity of β follows from

$$\beta([p]_\mathcal{D} + [q]_\mathcal{D}) = \beta([p \oplus q]_\mathcal{D}) = \nu(p \oplus q) = \nu(p) + \nu(q)$$
$$= \beta([p]_\mathcal{D}) + \beta([q]_\mathcal{D}).$$

Now use Paragraph 3.1.1 (i) to find a group homomorphism $\alpha \colon K_0(A) \to G$ that makes the diagram (3.3) commutative. The uniqueness of α follows from (3.2). □

3.2 Functoriality of K_0

3.2.1 Categories and functors. A category $\mathbf{C}$ consists of a class $\mathscr{O}(\mathbf{C})$ of *objects* and for each pair of objects A, B in $\mathscr{O}(\mathbf{C})$ a set $\mathrm{Mor}(A, B)$ of *morphisms* (from A to B) with an associative rule of composition

$$\mathrm{Mor}(A, B) \times \mathrm{Mor}(B, C) \to \mathrm{Mor}(A, C), \qquad (\varphi, \psi) \mapsto \psi \circ \varphi,$$

such that for each object A there is an element id_A in $\mathrm{Mor}(A, A)$ which satisfies $\mathrm{id}_B \circ \varphi = \varphi = \varphi \circ \mathrm{id}_A$ for every φ in $\mathrm{Mor}(A, B)$.

The two main examples of categories appearing in this text are the category $\mathbf{C^*}$-**alg** of C^*-algebras and the category $\mathbf{Ab}$ of Abelian groups. The class of objects, $\mathscr{O}(\mathbf{C^*}\text{-}\mathbf{alg})$, in $\mathbf{C^*}$-**alg** is the class of all C^*-algebras, and $\mathrm{Mor}(A, B)$ is the set of *-homomorphisms from A to B with the usual composition. The objects in $\mathbf{Ab}$ are Abelian groups, and morphisms are group homomorphisms. We shall in this chapter also consider the category of unital C^*-algebras, where the morphisms are *-homomorphisms (not necessarily unit preserving).

A *covariant functor* F between categories $\mathbf{C}$ and $\mathbf{D}$ is a map $A \mapsto F(A)$ from $\mathscr{O}(\mathbf{C})$ to $\mathscr{O}(\mathbf{D})$ and a collection of maps $\varphi \mapsto F(\varphi)$ from $\mathrm{Mor}(A, B)$ to $\mathrm{Mor}(F(A), F(B))$ for each pair of objects A, B in $\mathscr{O}(\mathbf{C})$ such that

(i) $F(\mathrm{id}_A) = \mathrm{id}_{F(A)}$ for all objects A in $\mathcal{O}(\mathbf{C})$,

(ii) $F(\psi \circ \varphi) = F(\psi) \circ F(\varphi)$ for all objects A, B, C in $\mathcal{O}(\mathbf{C})$ and all morphisms φ in $\mathrm{Mor}(A, B)$ and ψ in $\mathrm{Mor}(B, C)$.

A *contravariant functor* F is like a covariant functor with the exception that it reverses the direction of morphisms, i.e., it maps a morphism φ in $\mathrm{Mor}(A, B)$ to a morphism $F(\varphi)$ in $\mathrm{Mor}(F(B), F(A))$, and (ii) is changed accordingly. We shall in this text almost exclusively be concerned with covariant functors, and we refer to these for short as functors.

An object N in a category $\mathbf{C}$ is called a *zero object* if $\mathrm{Mor}(A, N)$ and $\mathrm{Mor}(N, A)$ contain precisely one element for all objects A in $\mathbf{C}$. (See also Exercise 3.9.) The zero C^*-algebra $\{0\}$ (also denoted by 0) is a zero object in $\mathbf{C^*}$-**alg**, and, similarly, the zero group is a zero object in $\mathbf{Ab}$. A functor F between categories with zero objects is said to preserve the zero objects if it maps zero objects to zero objects.

3.2.2 The functor K_0 for unital C^*-algebras. Let A and B be unital C^*-algebras, and let $\varphi \colon A \to B$ be a $*$-homomorphism. Associate to φ a group homomorphism $K_0(\varphi) \colon K_0(A) \to K_0(B)$ as follows. By Section 1.3, φ extends to a $*$-homomorphism $\varphi \colon M_n(A) \to M_n(B)$ for each n. A $*$-homomorphism maps projections to projections, and so φ maps $\mathcal{P}_\infty(A)$ into $\mathcal{P}_\infty(B)$. Define $\nu \colon \mathcal{P}_\infty(A) \to K_0(B)$ by $\nu(p) = [\varphi(p)]_0$ for p in $\mathcal{P}_\infty(A)$. Then ν satisfies (i), (ii), and (iii) in Proposition 3.1.8, and ν therefore factors uniquely through a group homomorphism $K_0(\varphi) \colon K_0(A) \to K_0(B)$ given by

$$(3.4) \qquad K_0(\varphi)([p]_0) = [\varphi(p)]_0, \qquad p \in \mathcal{P}_\infty(A).$$

In other words, we have a commutative diagram:

$$(3.5) \qquad
\begin{array}{ccc}
\mathcal{P}_\infty(A) & \xrightarrow{\;\varphi\;} & \mathcal{P}_\infty(B) \\
{\scriptstyle[\cdot]_0}\big\downarrow & & \big\downarrow{\scriptstyle[\cdot]_0} \\
K_0(A) & \xrightarrow[K_0(\varphi)]{} & K_0(B).
\end{array}$$

3.2.3 The functor K_{00}. If A and B are (not necessarily unital) C^*-algebras, then each $*$-homomorphism $\varphi \colon A \to B$ induces a group homomorphism $K_{00}(\varphi) \colon K_{00}(A) \to K_{00}(B)$ satisfying $K_{00}(\varphi)([p]_{00}) = [\varphi(p)]_{00}$ for each p in $\mathcal{P}_\infty(A)$. To see this, observe first that Proposition 3.1.8 can be extended word for word from K_0 to K_{00}, and then copy the argument in Paragraph 3.2.2.

If A and B are C^*-algebras, then denote the zero homomorphism $A \to B$ (that sends all elements in A to the zero element in B) by $0_{B,A}$ or just 0. (See also Exercise 3.9.) The identity map $A \to A$ is denoted by id_A. We shall use a similar notation in the category of Abelian groups.

Parts (i) and (ii) of the proposition below say that K_0 is a functor from the category of unital C^*-algebras to the category of Abelian groups. Parts (iii) and (iv) say that K_0 maps zero objects to zero objects and zero morphisms to zero morphisms. We include the zero C^*-algebra in the ranks of unital C^*-algebras!

Proposition 3.2.4 (Functoriality of K_0 for unital C^*-algebras).

(i) *For each unital C^*-algebra A, $K_0(\mathrm{id}_A) = \mathrm{id}_{K_0(A)}$.*

(ii) *If A, B, and C are unital C^*-algebras, and if $\varphi: A \to B$ and $\psi: B \to C$ are *-homomorphisms, then $K_0(\psi \circ \varphi) = K_0(\psi) \circ K_0(\varphi)$.*

(iii) $K_0(\{0\}) = \{0\}$.

(iv) *For every pair of C^*-algebras A and B, $K_0(0_{B,A}) = 0_{K_0(B),K_0(A)}$.*

Proof. Use (3.4) to check that

$$K_0(\mathrm{id}_A)([p]_0) = [p]_0, \qquad K_0(\psi \circ \varphi)([p]_0) = (K_0(\psi) \circ K_0(\varphi))([p]_0),$$

for every p in $\mathcal{P}_\infty(A)$. Then use the standard picture of K_0 (equation (3.2)) to conclude that (i) and (ii) hold.

(iii). We have $\mathcal{P}_n(\{0\}) = \{0_n\}$, where 0_n is the zero element in $M_n(\{0\})$. The zero projections $0 = 0_1, 0_2, \ldots$ are all equivalent, and so $\mathcal{D}(\{0\}) = \{[0]_\mathcal{D}\}$. Hence $K_0(\{0\}) = G(\{0\}) = \{0\}$.

(iv). Since $0_{B,A} = 0_{B,0} \circ 0_{0,A}: A \to \{0\} \to B$, (iv) follows from (ii) and (iii). $\square$

A similar argument shows that K_{00} is a functor that preserves the zero objects.

Definition 3.2.5 (Homotopy equivalence). Let A and B be C^*-algebras. Two *-homomorphisms $\varphi, \psi: A \to B$ are said to be *homotopic*, in symbols $\varphi \sim_h \psi$, if there is a path of *-homomorphisms $\varphi_t: A \to B$, $t \in [0,1]$, such that $t \mapsto \varphi_t(a)$ is a continuous map from $[0,1]$ to B for each a in A, $\varphi_0 = \varphi$, and $\varphi_1 = \psi$. We say that the path $t \mapsto \varphi_t$ is *point-wise continuous*.

The C^*-algebras A and B are *homotopy equivalent* if there are *-homomorphisms $\varphi\colon A \to B$ and $\psi\colon B \to A$ such that $\psi \circ \varphi \sim_h \mathrm{id}_A$ and $\varphi \circ \psi \sim_h \mathrm{id}_B$. In this case we say that

$$A \xrightarrow{\varphi} B \xrightarrow{\psi} A$$

is a *homotopy* (between A and B).

Proposition 3.2.6 (Homotopy invariance of K_0). *Let A and B be unital C^*-algebras.*

(i) *If $\varphi, \psi\colon A \to B$ are homotopic *-homomorphisms, then $K_0(\varphi) = K_0(\psi)$.*

(ii) *If A and B are homotopy equivalent, then $K_0(A)$ is isomorphic to $K_0(B)$. More specifically, if*

$$A \xrightarrow{\varphi} B \xrightarrow{\psi} A$$

is a homotopy, then $K_0(\varphi)\colon K_0(A) \to K_0(B)$ and $K_0(\psi)\colon K_0(B) \to K_0(A)$ are isomorphisms, and $K_0(\varphi)^{-1} = K_0(\psi)$.

Proof. (i). Let $\varphi_t\colon A \to B$ be a point-wise continuous path of *-homomorphisms connecting φ to ψ. Extend this path to a point-wise continuous path of *-homomorphisms $\varphi_t\colon M_n(A) \to M_n(B)$ for each n in $\mathbb{N}$ (see Section 1.3). For every projection p in $\mathcal{P}_n(A)$, the path $t \mapsto \varphi_t(p)$ is continuous, and so $\varphi(p) = \varphi_0(p) \sim_h \varphi_1(p) = \psi(p)$. This shows that

$$K_0(\varphi)([p]_0) = [\varphi(p)]_0 = [\psi(p)]_0 = K_0(\psi)([p]_0).$$

Use the standard picture of K_0 (equation (3.2)) to conclude that $K_0(\varphi) = K_0(\psi)$.

Part (ii) follows from (i) and from Proposition 3.2.4 (i) and (ii). $\quad\square$

We proceed to show that K_0 preserves exactness of the short exact sequence obtained by adjoining a unit to a unital C^*-algebra. This result will be useful when we define K_0 for non-unital C^*-algebras.

Two *-homomorphisms $\varphi, \psi\colon A \to B$ between C^*-algebras A and B are said to be *orthogonal to each other*, or mutually orthogonal, in symbols, $\varphi \perp \psi$, if $\varphi(x)\psi(y) = 0$ for all x, y in A.

Lemma 3.2.7. *If A and B are unital C^*-algebras, and if $\varphi, \psi\colon A \to B$ are mutually orthogonal *-homomorphisms, then $\varphi + \psi\colon A \to B$ is a *-homomorphism, and $K_0(\varphi + \psi) = K_0(\varphi) + K_0(\psi)$.*

Proof. It is straightforward to check that $\varphi + \psi$ is a *-homomorphism. The
*-homomorphisms $\varphi_n, \psi_n \colon M_n(A) \to M_n(B)$ induced by φ and ψ (see Section 1.3) are orthogonal to each other for each n in $\mathbb{N}$, and $(\varphi + \psi)_n = \varphi_n + \psi_n$.
Using the standard picture of K_0 (Proposition 3.1.7 (iv)) we obtain for every
p in $\mathcal{P}_n(A)$

$$
\begin{aligned}
K_0(\varphi + \psi)([p]_0) &= [(\varphi + \psi)_n(p)]_0 = [\varphi_n(p) + \psi_n(p)]_0 \\
&= [\varphi_n(p)]_0 + [\psi_n(p)]_0 \\
&= K_0(\varphi)([p]_0) + K_0(\psi)([p]_0).
\end{aligned}
$$

This shows that $K_0(\varphi + \psi) = K_0(\varphi) + K_0(\psi)$. $\qquad\square$

Lemma 3.2.8. *For every unital C^*-algebra A, the split exact sequence*

$$
0 \longrightarrow A \overset{\iota}{\longrightarrow} \widetilde{A} \underset{\lambda}{\overset{\pi}{\rightleftarrows}} \mathbb{C} \longrightarrow 0,
$$

obtained by adjoining a unit to A, induces a split exact sequence

$$
(3.6) \qquad 0 \longrightarrow K_0(A) \xrightarrow{\;K_0(\iota)\;} K_0(\widetilde{A}) \underset{K_0(\lambda)}{\overset{K_0(\pi)}{\rightleftarrows}} K_0(\mathbb{C}) \longrightarrow 0.
$$

Proof. Recall from the construction of adjoining a unit to a (unital) C^*-algebra that if f is the projection $1_{\widetilde{A}} - 1_A$ in $\widetilde{A}$, then $\widetilde{A} = A + \mathbb{C}f$ and $af = fa = 0$
for all a in A. Define *-homomorphisms

$$
\mu \colon \widetilde{A} \to A, \qquad \lambda' \colon \mathbb{C} \to \widetilde{A}
$$

by $\mu(a + \alpha f) = a$ and $\lambda'(\alpha) = \alpha f$. Then

$$
\mathrm{id}_A = \mu \circ \iota, \quad \mathrm{id}_{\widetilde{A}} = \iota \circ \mu + \lambda' \circ \pi, \quad \pi \circ \iota = 0, \quad \pi \circ \lambda = \mathrm{id}_{\mathbb{C}},
$$

and the *-homomorphisms $\iota \circ \mu$ and $\lambda' \circ \pi$ are orthogonal to each other.
Functoriality of K_0 (Proposition 3.2.4) and Lemma 3.2.7 now yield

$$
\begin{aligned}
0 &= K_0(0) = K_0(\pi \circ \iota) = K_0(\pi) \circ K_0(\iota), \\
\mathrm{id}_{K_0(\mathbb{C})} &= K_0(\mathrm{id}_{\mathbb{C}}) = K_0(\pi \circ \lambda) = K_0(\pi) \circ K_0(\lambda), \\
\mathrm{id}_{K_0(A)} &= K_0(\mathrm{id}_A) = K_0(\mu \circ \iota) = K_0(\mu) \circ K_0(\iota), \\
\mathrm{id}_{K_0(\widetilde{A})} &= K_0(\mathrm{id}_{\widetilde{A}}) = K_0(\iota \circ \mu + \lambda' \circ \pi) \\
&= K_0(\iota) \circ K_0(\mu) + K_0(\lambda') \circ K_0(\pi).
\end{aligned}
$$

Split exactness of (3.6) follows from these four identities: Injectivity of $K_0(\iota)$
follows from the third identity. If g belongs to the kernel of $K_0(\pi)$, then

$g = K_0(\iota)(K_0(\mu)(g))$ by the last identity, thus showing that g belongs to the image of $K_0(\iota)$.   □

3.3   Examples

3.3.1 Traces and K_0. Let A be a C^*-algebra. A bounded *trace* on A is a bounded linear map $\tau: A \to \mathbb{C}$ with the *trace property*:

(3.7) $\tau(ab) = \tau(ba), \qquad a, b \in A.$

(See also Exercise 3.6.) The trace property implies that $\tau(p) = \tau(q)$ whenever p, q are Murray–von Neumann equivalent projections in A.

A trace τ is *positive* if $\tau(a) \geqslant 0$ for every positive element a in A. If A is unital and τ is a positive trace with $\tau(1_A) = 1$, then τ is called a *tracial state*, see Paragraph 1.2.2.

For every trace τ on a C^*-algebra A there is precisely one trace τ_n (usually abbreviated to τ) on $M_n(A)$ that satisfies $\tau_n(\mathrm{diag}(a, 0, \ldots, 0)) = \tau(a)$ for all a in A, and τ_n is given by

(3.8)
$$\tau_n \begin{pmatrix} a_{11} & a_{12} & \cdots & a_{1n} \\ a_{21} & a_{22} & \cdots & a_{2n} \\ \vdots & \vdots & \ddots & \vdots \\ a_{n1} & a_{n2} & \cdots & a_{nn} \end{pmatrix} = \sum_{i=1}^{n} \tau(a_{ii}).$$

(See Exercise 3.5.)

A trace τ on a C^*-algebra A gives rise in this way to a function $\tau: \mathcal{P}_\infty(A) \to \mathbb{C}$, and this function satisfies conditions (i), (ii), and (iii) in Proposition 3.1.8 (the universal property of K_0), and so there is a unique group homomorphism $K_0(\tau): K_0(A) \to \mathbb{C}$ satisfying

(3.9) $K_0(\tau)([p]_0) = \tau(p), \qquad p \in \mathcal{P}_\infty(A).$

To see that (iii) in Proposition 3.1.8 holds, use that $p \sim q$ if $p \sim_h q$. If τ is positive, then $K_0(\tau)([p]_0) = \tau(p)$ is a positive real number for every p in $\mathcal{P}_\infty(A)$, and $K_0(\tau)$ maps $K_0(A)$ into $\mathbb{R}$.

Example 3.3.2. The group $K_0(M_n(\mathbb{C}))$ is isomorphic to $\mathbb{Z}$ for each positive integer n. More specifically, if Tr is the standard trace on $M_n(\mathbb{C})$, (see Exercise 2.9), then

$$K_0(\mathrm{Tr})\colon K_0(M_n(\mathbb{C})) \to \mathbb{Z}$$

is an isomorphism. The cyclic group $K_0(M_n(\mathbb{C}))$ is generated by $[e]_0$, where e is any one-dimensional projection in $M_n(\mathbb{C})$.

Since $\mathbb{C} = M_1(\mathbb{C})$ we obtain in particular that $K_0(\mathbb{C})$ is isomorphic to $\mathbb{Z}$.

Proof. Let g be an element in $K_0(M_n(\mathbb{C}))$. By the standard picture of K_0 (equation (3.2)) we can find k in $\mathbb{N}$ and projections p, q in $M_k(M_n(\mathbb{C})) = M_{kn}(\mathbb{C})$ such that $g = [p]_0 - [q]_0$. Now,

$$K_0(\mathrm{Tr})(g) = \mathrm{Tr}(p) - \mathrm{Tr}(q) = \dim(p(\mathbb{C}^{kn})) - \dim(q(\mathbb{C}^{kn})).$$

We see that $K_0(\mathrm{Tr})(g)$ is an integer. If $K_0(\mathrm{Tr})(g) = 0$, then $p \sim q$ by Exercise 2.9, whence $g = [p]_0 - [q]_0 = 0$. Hence $K_0(\mathrm{Tr})$ is injective.

The image of $K_0(\mathrm{Tr})$ is a subgroup of $\mathbb{Z}$, and a subgroup of $\mathbb{Z}$ is equal to $\mathbb{Z}$ if and only if it contains 1. Now, $1 = K_0(\mathrm{Tr})([e]_0)$, when e is a projection in $M_n(\mathbb{C})$ with one-dimensional range. $\qquad\square$

Example 3.3.3. Let H be an infinite dimensional Hilbert space. Then we have $K_0(B(H)) = 0$.

Proof. We first consider the case where H is an infinite dimensional, separable Hilbert space. Let H^n denote the Hilbert space $H \oplus H \oplus \cdots \oplus H$ (with n summands). We use the identification $M_n(B(H)) = B(H^n)$ (with a slight abuse of notation). The map $\dim\colon \mathcal{P}_\infty(B(H)) \to \{0, 1, 2, \ldots, \infty\}$ given by

$$\dim(p) = \dim(p(H^n)), \qquad p \in \mathcal{P}_n(B(H)) = \mathcal{P}(B(H^n)),$$

is surjective. It is shown in Exercise 2.10 that if p, q are projections in $\mathcal{P}(B(H^n))$, then $\dim(p) = \dim(q)$ if and only if $p \sim q$. Since $\dim(p \oplus 0) = \dim(p)$, we obtain for all projections p, q in $\mathcal{P}_\infty(B(H))$ that $\dim(p) = \dim(q)$ if and only if $p \sim_0 q$. Dimension is additive, and so $\dim(p \oplus q) = \dim(p) + \dim(q)$.

This shows that $d([p]_{\mathcal{D}}) = \dim(p)$ is a well-defined semigroup isomorphism $d\colon \mathcal{D}(B(H)) \to \{0, 1, 2, \ldots, \infty\}$. We conclude that $K_0(B(H))$ is isomorphic to the Grothendieck group of the semigroup $\{0, 1, 2, \ldots, \infty\}$, and by Example 3.1.3 (ii) this Grothendieck group is zero, thus showing that $K_0(B(H)) = 0$.

If H is not separable, then $\mathcal{D}(B(H))$ is isomorphic to the semigroup of all cardinal numbers less than or equal to $\dim(H)$ (where $\dim(H)$ is the cardinality of an orthonormal basis for H). The Grothendieck group of any such semigroup is zero, and hence $K_0(B(H)) = 0$ also in the non-separable case. □

Example 3.3.4. If X is a connected, locally compact, non-compact Hausdorff space, then $K_{00}(C_0(X)) = 0$, see Paragraph 3.1.5.

Proof. We may identify $M_n(C_0(X))$ with $C_0(X, M_n(\mathbb{C}))$. Under this identification $\mathcal{P}_n(C_0(X))$ is identified with $\mathcal{P}(C_0(X, M_n(\mathbb{C})))$. Let p be a projection in $\mathcal{P}_n(C_0(X))$. As usual, let Tr denote the standard trace on $M_n(\mathbb{C})$. The function $x \mapsto \mathrm{Tr}(p(x))$ belongs to $C_0(X, \mathbb{Z})$. But $C_0(X, \mathbb{Z}) = \{0\}$ because X is connected and non-compact. Therefore $\mathcal{P}_n(C_0(X)) = \{0\}$ for all natural numbers n, and this shows that $K_{00}(C_0(X)) = 0$. □

Example 3.3.5. For every compact, connected Hausdorff space X there is a surjective group homomorphism

$$\mathrm{dim}\colon K_0(C(X)) \to \mathbb{Z}$$

that satisfies

$$\mathrm{dim}([p]_0) = \mathrm{Tr}(p(x)), \qquad p \in \mathcal{P}_\infty(C(X)),$$

where x is any element in X, and where Tr is the standard trace on $M_n(\mathbb{C})$.

Proof. We begin by showing that $\mathrm{Tr}(p(x))$ is independent of x. The function $x \mapsto \mathrm{Tr}(p(x))$ belongs to $C(X, \mathbb{Z})$. Since X is connected, every function in $C(X, \mathbb{Z})$ is constant. Therefore the map $x \mapsto \mathrm{Tr}(p(x))$ is constant for each p in $\mathcal{P}_\infty(C(X))$.

The map $\tau_x\colon C(X) \to \mathbb{C}$ given by $\tau_x(f) = f(x)$ is a trace on $C(X)$ for each x in X. Hence τ_x induces a group homomorphism $K_0(\tau_x)\colon K_0(C(X)) \to \mathbb{C}$ which satisfies $K_0(\tau_x)([p]_0) = \tau_x(p)$ for each projection p in $\mathcal{P}_\infty(C(X))$. If p belongs to $\mathcal{P}_n(C(X))$, then $\tau_x(p) = \mathrm{Tr}(p(x))$ by the convention on how a trace is extended to matrix algebras. We conclude that the homomorphism $K_0(\tau_x)$ is independent of x and that it has range contained in $\mathbb{Z}$ because $\mathrm{Tr}(p(x))$ is an integer for every p in $\mathcal{P}_\infty(C(X))$. Setting $\mathrm{dim} = K_0(\tau_x)$, we need only show that $K_0(\tau_x)$ is surjective.

The unit 1 in $C(X)$ is a projection, and $1 = 1(x) = K_0(\tau_x)([1]_0)$. This entails that $K_0(\tau_x)$ is surjective. □

Example 3.3.6. A compact Hausdorff space X is called *contractible* if, for some element x_0 in X, there is a continuous map $\alpha\colon [0,1] \times X \to X$, such that $\alpha(1, x) = x$ and $\alpha(0, x) = x_0$ for all x in X.

Let X be a contractible, compact Hausdorff space. Then $K_0(C(X))$ is isomorphic to $\mathbb{Z}$ and the map $\dim\colon K_0(C(X)) \to \mathbb{Z}$ from Example 3.3.5 is an isomorphism.

Proof. Assume that X is contractible, and let x_0 and α be as above. Define for each t in $[0,1]$ a *-homomorphism $\varphi_t\colon C(X) \to C(X)$ by $\varphi_t(f)(x) = f(\alpha(t,x))$. Then $\varphi_0(f)(x) = f(x_0)$ and $\varphi_1 = \mathrm{id}$. Moreover, the map $t \mapsto \varphi_t(f)$ is continuous for each f in $C(X)$ (see Exercise 3.13). This shows that $\varphi_0 \sim_h \mathrm{id}$.

Define $\varphi\colon C(X) \to \mathbb{C}$ and $\psi\colon \mathbb{C} \to C(X)$ by $\varphi(f) = f(x_0)$ and $\psi(\lambda) = \lambda{\cdot}1$. Then $\varphi \circ \psi = \mathrm{id}_{\mathbb{C}}$ and $\psi \circ \varphi = \varphi_0 \sim_h \mathrm{id}$. Hence

$$C(X) \xrightarrow{\;\varphi\;} \mathbb{C} \xrightarrow{\;\psi\;} C(X)$$

is a homotopy. The diagram

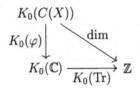

is commutative, and $K_0(\varphi)$ and $K_0(\mathrm{Tr})$ are isomorphisms by homotopy invariance of K_0 (Proposition 3.2.6) and by Example 3.3.2, thus forcing $\dim$ to be an isomorphism.   □

See Exercise 3.3 for a couple of (trivial) examples of contractible spaces.

3.3.7* Topological K-theory of spaces. The topological K-theory of a topological space X is the ring of isomorphism classes of (complex) *vector bundles* over X with addition given by direct sum of vector bundles and product given by tensor product of vector bundles. We give below a brief description of this theory with emphasis on how it connects to K-theory for C^*-algebras. The reader is referred to Atiyah's book [1], or to Husemoller [25], for more information.

A complex vector bundle over a topological space X is a triple $\xi = (E, \pi, X)$, where E is a topological space, $\pi\colon E \to X$ is a continuous surjection, and where for each x in X the fiber $E_x = \pi^{-1}(x)$ is equipped with the structure of a finite dimensional complex vector space, such that ξ is *locally trivial* in the following sense: ξ is said to be trivial over an open subset

U of X if there exist n in $\mathbb{N}$ and a homeomorphism $h\colon \pi^{-1}(U) \to U \times \mathbb{C}^n$ making the diagram

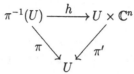

commutative, where $\pi'(x, v) = x$, and such that for each x in U the restriction of h to E_x yields a vector space isomorphism $E_x \to \{x\} \times \mathbb{C}^n \to \mathbb{C}^n$. If each point x in X has an open neighborhood over which ξ is trivial, then ξ is said to be locally trivial.

An example of a vector bundle is the triple $\theta_n = (X \times \mathbb{C}^n, \pi, X)$, where π is the coordinate mapping $\pi(x, v) = x$. This vector bundle is called the *trivial vector bundle* over X of dimension n. If $\xi = (E, \pi, X)$ is a vector bundle such that each fiber E_x has dimension n, then n is said to be the dimension of ξ. If X is disconnected, then different fibers in a vector bundle over X need not have the same dimension.

Two vector bundles $\xi = (E, \pi, X)$ and $\eta = (F, \rho, X)$ over X are *isomorphic*, written $\xi \cong \eta$, if there is a homeomorphism $h\colon E \to F$ making the diagram

commutative, and such that the restriction of h to E_x is a vector space isomorphism $E_x \to F_x$ for each x in X. Let $\langle \xi \rangle$ denote the class of all vector bundles over X that are isomorphic to ξ.

A vector bundle ξ over X is said to be *trivial* if it is trivial over the (open) set X. Equivalently, ξ is trivial if and only if ξ is isomorphic to θ_n for some n.

The direct sum $\xi \oplus \eta$ of two vector bundles $\xi = (E, \pi, X)$ and $\eta = (F, \rho, X)$ over X is defined to be the triple (G, ν, X), where

$$G = \{(v, w) \in E \times F : \pi(v) = \rho(w)\},$$

$\nu(v, w) = \pi(v) = \rho(w)$, and where the fibers $G_x = E_x \times F_x$ are given the direct sum vector space structure. One can define the (interior) *tensor product* $\xi \otimes \eta$ of two vector bundles ξ and η over X to be the vector bundle whose fiber over x is the tensor product of the fiber of ξ over x and the fiber of η over x. (See [1] or [25] for more details.)

Define $\text{Vect}(X)$ to be the set of all isomorphism classes $\langle \xi \rangle$ of vector bundles ξ over X. Define addition and multiplication on $\text{Vect}(X)$ by $\langle \xi \rangle + \langle \eta \rangle = \langle \xi \oplus \eta \rangle$ and $\langle \xi \rangle \cdot \langle \eta \rangle = \langle \xi \otimes \eta \rangle$. These operations are well-defined, $(\text{Vect}(X), +)$ is an Abelian semigroup, multiplication is commutative, and the distributive law holds.

Define $K^0(X)$ to be the Grothendieck group of $(\text{Vect}(X), +)$, and extend the multiplication on $\text{Vect}(X)$ to a multiplication on $K^0(X)$. Let $[\xi]$ denote the image of $\langle \xi \rangle$ in $K^0(X)$ under the Grothendieck map $\text{Vect}(X) \to K^0(X)$. Then $K^0(X)$ becomes a commutative ring with unit $[\theta_1]$. It is common to write $K(X)$ instead of $K^0(X)$.

Let Y be another topological space, let $f \colon X \to Y$ be a continuous function, and let $\eta = (F, \rho, Y)$ be a vector bundle over Y. Define $f_*\eta$ to be the triple (E, π, X), where

$$E = \{ (x, v) \ \epsilon \ X \times F : f(x) = \rho(v) \},$$

$\pi(x, v) = x$, and where the fiber $E_x = F_{f(x)}$ is given the vector space structure from η. The function f induces a map $f_* \colon \text{Vect}(Y) \to \text{Vect}(X)$ by $f_*(\langle \eta \rangle) = \langle f_*\eta \rangle$ and a map $K(f) \colon K(Y) \to K(X)$ that satisfies $K(f)([\xi]) = [f_*\xi]$. The map $K(f)$ is a unit preserving ring homomorphism. In this way K becomes a contravariant functor from the category of topological spaces to the category of unital rings. (The upper index on the K indicates that the functor is contravariant, whereas the lower index on C^*-algebra K-theory indicates that this functor is covariant.)

We proceed to show that if X is a compact Hausdorff space, then $K_0(C(X))$ and $K(X)$ are isomorphic as Abelian groups, and what is more, the semigroups $\mathcal{D}(C(X))$ and $\text{Vect}(X)$ are isomorphic. To see this, associate a vector bundle ξ_p to each projection p in $\mathcal{P}_\infty(C(X))$ as follows. Suppose that p belongs to $\mathcal{P}_n(C(X))$, and set $\xi_p = (E_p, \pi, X)$, where

$$E_p = \{ (x, v) \ \epsilon \ X \times \mathbb{C}^n : v \ \epsilon \ p(x)(\mathbb{C}^n) \},$$

using the identification $M_n(C(X)) = C(X, M_n(\mathbb{C}))$, $\pi(x, v) = x$, and where the fiber $(E_p)_x = p(x)(\mathbb{C}^n)$ is given the vector space structure making it a subspace of $\mathbb{C}^n$. If p, q belong to $\mathcal{P}_\infty(C(X))$, then $\xi_p \cong \xi_q$ if and only if $p \sim_0 q$, and $\xi_{p \oplus q} \cong \xi_p \oplus \xi_q$. It follows that the map $\mathcal{D}(C(X)) \to \text{Vect}(X)$ given by $[p]_\mathcal{D} \mapsto \langle \xi_p \rangle$ is a well-defined, injective, and additive map. A theorem of R. G. Swan ([39]) says that each vector bundle $\xi = (E, \pi, X)$ is a direct summand in a trivial bundle, i.e., there exist n and a vector bundle $\eta =$

(F, ρ, X) such that $\xi \oplus \eta \cong \theta_n$. We may therefore assume that $E_x \oplus F_x = \mathbb{C}^n$ for each x in X. Let $p(x)$ be the projection from $\mathbb{C}^n$ onto E_x. Then $p \colon X \to M_n(\mathbb{C})$ is continuous, so that p belongs to $\mathcal{P}_n(C(X))$, and $\xi_p = \xi$. This shows that the map $\mathcal{D}(C(X)) \to \mathrm{Vect}(X)$ is surjective, and hence an isomorphism of semigroups.

There is no natural product on $K_0(A)$ when A is a C^*-algebra unless when $A = C(X)$. A further structure of topological K-theory for spaces is found in the *Chern character*, ch: $K^0(X) \to H^{ev}(X; \mathbb{Q})$, which is a ring homomorphism from the K-theory of X into the even cohomology of X with rational coefficients, and which induces a ring *isomorphism* $K^0(X) \otimes \mathbb{Q} \to H^{ev}(X; \mathbb{Q})$. The Chern character has been extended to a non-commutative Chern character into the *cyclic cohomology* of algebras (with a smooth structure) by A. Connes (see [11]).

3.3.8 Exactness properties of functors. It is a central issue in K-theory to what extent a functor (in particular a K-functor) from the category of C^*-algebras to the category of Abelian groups (or between other categories) preserves short exact sequences.

Let F be a functor from the category of C^*-algebras to the category of Abelian groups that preserves zero objects, i.e., $F(\{0\}) = \{0\}$. Then each short exact sequence

$$(3.10) \qquad 0 \longrightarrow I \xrightarrow{\;\varphi\;} A \xrightarrow{\;\psi\;} B \longrightarrow 0$$

of C^*-algebras induces a (not necessarily exact) sequence

$$(3.11) \qquad 0 \longrightarrow F(I) \xrightarrow{\;F(\varphi)\;} F(A) \xrightarrow{\;F(\psi)\;} F(B) \longrightarrow 0$$

of Abelian groups. If the sequence (3.11) is exact for all short exact sequences (3.10), then F is called *exact*. If the sequence (3.11) is split exact for all split exact sequences (3.10), then F is called *split exact*. If the sequence (3.11) is exact at $F(A)$, i.e., if $\mathrm{Im}(F(\varphi)) = \mathrm{Ker}(F(\psi))$ for all short exact sequences (3.10), then F is called *half exact*.

Example 3.3.9* (K_{00} is not half exact). It is shown in this example that the functor K_{00} defined in Paragraph 3.1.5 is not half exact.

Let X be a connected, compact Hausdorff space, and choose any point x_0

in X. Then we get a split exact sequence

$$0 \longrightarrow C_0(X \setminus \{x_0\}) \xrightarrow{\iota} C(X) \underset{\lambda}{\overset{\pi}{\rightleftarrows}} \mathbb{C} \longrightarrow 0,$$

where $\pi(f) = f(x_0)$, and where the lift λ maps a scalar α to the constant function α. We know from the proof of Example 3.3.6 that the diagram

(3.12)

$$\begin{array}{c} K_0\big(C(X)\big) \\ K_0(\pi) \Big\downarrow \quad \diagdown \ \text{dim} \\ K_0(\mathbb{C}) \xrightarrow[K_0(\text{Tr})]{} \mathbb{Z} \end{array}$$

is commutative, and $K_0(\text{Tr})$ is an isomorphism by Example 3.3.2. Hence $\text{Ker(dim)} = \text{Ker}(K_0(\pi))$.

Next, $X \setminus \{x_0\}$ is a non-compact, locally compact Hausdorff space. If $X\setminus\{x_0\}$ is connected, then Example 3.3.5 tells us that $K_{00}(C_0(X\setminus\{x_0\})) = 0$. (It is not hard to show that $K_{00}(C_0(X\setminus\{x_0\})) = 0$ also when $X\setminus\{x_0\}$ is not a connected space, but let us avoid this argument for now.) Assuming that $X\setminus\{x_0\}$ is connected, we find that the sequence

(3.13)

$$\begin{array}{ccc} K_{00}(C_0(X\setminus\{x_0\})) \xrightarrow{K_{00}(\iota)} K_{00}(C(X)) \xrightarrow{K_{00}(\pi)} K_{00}(\mathbb{C}) \\ \Big\| \qquad\qquad\qquad \Big\| \\ K_0(C(X)) \xrightarrow[K_0(\pi)]{} K_0(\mathbb{C}) \end{array}$$

is exact at $K_{00}(C(X))$ if and only if the map dim: $K_0(C(X)) \to \mathbb{Z}$ in (3.12) is an isomorphism. In other words, to show that K_{00} is not half exact, we need only exhibit a connected, compact Hausdorff space X for which dim: $K_0(C(X)) \to \mathbb{Z}$ is not injective.

An example of such a space is $X = S^2$. It will be shown in Example 11.3.3 that $K_0(C(S^2))$ is isomorphic to $\mathbb{Z} \oplus \mathbb{Z}$, and this will entail that the mapping

$$\text{dim}: K_0(C(S^2)) \to \mathbb{Z}$$

cannot be injective. (See also Exercise 9.5.)

We can at this point give a partial argument that perhaps makes it plausible that dim: $K_0(C(S^2)) \to \mathbb{Z}$ is not injective.

Identify $M_2(C(S^2))$ with $C(S^2, M_2(\mathbb{C}))$. The topological space $G_{2,1}$ of all one-dimensional projections in $M_2(\mathbb{C})$ is homeomorphic to S^2. (See Exercise 3.8.) Let e be a homeomorphism from S^2 onto $G_{2,1}$, and view e as an element in $C(S^2, M_2(\mathbb{C}))$. Then e is a projection, and e is not homotopic to the projection $f = \mathrm{diag}(1,0)$ in $\mathcal{P}_2(C(S^2))$. Indeed, if $t \mapsto p_t$ were a continuous path of projections in $\mathcal{P}_2(C(S^2))$ with $p_0 = e$ and $p_1 = f$, then $t \mapsto e^{-1} \circ p_t$ would be a continuous path of continuous functions $S^2 \to S^2$ connecting the identity map to a constant map — but the identity map on S^2 is not homotopic to a constant map.

It will be shown in detail in Exercise 9.5 that $[e]_0 - [f]_0 \neq 0$. Hence $[e]_0 - [f]_0$ is a non-zero element in the kernel of the map dim: $K_0(C(S^2)) \to \mathbb{Z}$.

3.4   Exercises

Exercise 3.1. Show that $K_0(A)$ is a *countable* Abelian group if A is a separable unital C^*-algebra. Show also that $K_{00}(A)$ is a countable Abelian group when A is any separable C^*-algebra. [Hint: Use Proposition 2.2.4.]

Exercise 3.2. Let A be a C^*-algebra, let G be an Abelian group, and let $\nu\colon \mathcal{P}_\infty(A) \to G$ be a map that satisfies $\nu(0_A) = 0$ and $\nu(p \oplus q) = \nu(p) + \nu(q)$ for all projections p, q in $\mathcal{P}_\infty(A)$. Show that the following four conditions are equivalent:

(i) for all n and for all projections p, q in $\mathcal{P}_n(A)$, if $p \sim_h q$, then $\nu(p) = \nu(q)$,

(ii) for all n and for all projections p, q in $\mathcal{P}_n(A)$, if $p \sim_u q$, then $\nu(p) = \nu(q)$,

(iii) for all projections p, q in $\mathcal{P}_\infty(A)$, if $p \sim_0 q$, then $\nu(p) = \nu(q)$,

(iv) for all projections p, q in $\mathcal{P}_\infty(A)$, if $p \sim_s q$, then $\nu(p) = \nu(q)$.

Exercise 3.3. Show that $[0,1]$ and $\mathbb{D}$ are contractible spaces (see Example 3.3.6), where $\mathbb{D}$ is the closed unit disc in the complex plane, and conclude that $K_0(C([0,1])) \cong \mathbb{Z}$ and $K_0(C(\mathbb{D})) \cong \mathbb{Z}$.

Exercise 3.4. Let X be any compact Hausdorff space.

(i) By generalizing Example 3.3.5, show that there is a surjective group homomorphism
$$\mathrm{dim}\colon K_0(C(X)) \to C(X, \mathbb{Z})$$
which satisfies $\mathrm{dim}([p]_0)(x) = \mathrm{Tr}(p(x))$.

(ii) Let p in $\mathcal{P}_n(C(X))$ and q in $\mathcal{P}_m(C(X))$ be given. Show that $\dim([p]_0) = \dim([q]_0)$ if and only if for each x in X there exists v_x in $M_{m,n}(C(X))$ such that $v_x^* v_x = p(x)$ and $v_x v_x^* = q(x)$. Note that one cannot in general choose v_x such that the map $x \mapsto v_x$ is continuous!

(iii) Show that the map dim in (i) is injective (and hence an isomorphism) if X is *totally disconnected*. A compact Hausdorff space is totally disconnected if it admits a basis for its topology consisting of *clopen* sets, i.e., sets that are both closed and open. [Hint: Use (ii) and the total disconnectedness of X to find a partition of X into clopen sets $X_1, X_2, \ldots, X_r$ and rectangular matrices $v_1, v_2, \ldots, v_r$ over $C(X)$ such that $\|v_i^* v_i - p(x)\| < 1$ and $\|v_i v_i^* - q(x)\| < 1$ for all x in X_i. Use this to show that $p \sim_0 q$.]

Exercise 3.5. Let A be a unital C^*-algebra, and let τ be a trace on A.

(i) Verify that the map $\tau_n\colon M_n(A) \to \mathbb{C}$ given in Paragraph 3.3.1 defines a trace on $M_n(A)$.

(ii) Assume that $\rho\colon M_n(A) \to \mathbb{C}$ is a trace and that $\rho(\mathrm{diag}(a,0,\ldots,0)) = \tau(a)$ for all a in A. Show that $\rho = \tau_n$. [Hint: Let e_{ij} be the matrix in $M_n(A)$ that has 1 in the (i,j)th entry and zeros elsewhere. Observe that $e_{ij}e_{jk} = e_{ik}$, that $e_{ij}e_{lk} = 0$ when $j \neq l$, and that $\sum_{i=1}^n e_{ii} = 1_n$. Observe also that $\rho(e_{1i}be_{j1}) = \tau_n(e_{1i}be_{j1})$ for all b in $M_n(\mathbb{C})$ and for all i,j. Show now that $\rho(e_{ii}be_{ii}) = \rho(e_{1i}be_{i1})$ and that $\rho(e_{ii}be_{jj}) = 0$ when $i \neq j$. Use this to show that $\rho = \tau_n$.]

(iii) Use (ii) to show that there is one and only one trace Tr on $M_n(\mathbb{C})$ such that $\mathrm{Tr}(e) = 1$, where $e = \mathrm{diag}(1,0,\ldots,0)$.

Exercise 3.6. Let $\tau\colon A \to \mathbb{C}$ be a linear functional on a C^*-algebra A. Show that the following conditions are equivalent:

(i) τ is a trace;

(ii) $\tau(x^*x) = \tau(xx^*)$ for all x in A;

(iii) $\tau(uau^*) = \tau(a)$ for all unitary u in $\widetilde{A}$ and for all positive elements a in A.

[Hint: Show (i) $\Rightarrow$ (ii) $\Rightarrow$ (iii) $\Rightarrow$ (i). To see that (iii) $\Rightarrow$ (i), show first that $\tau(ux) = \tau(xu)$ for all unitary u in $\widetilde{A}$ and for all x in A (see Exercise 1.7). Then use Exercise 2.2.]

Exercise 3.7. Let A and B be C^*-algebras, and assume that we are given a *-homomorphism $\varphi_t \colon A \to B$ for each t in $[0,1]$. Show that the set of elements a in A for which the function $t \mapsto \varphi_t(a)$ is continuous is a sub-C^*-algebra of A.

   Suppose that F is a generating subset of A, in other words, $A = C^*(F)$ (see Paragraph 1.1.2), and that the function $t \mapsto \varphi_t(x)$ is continuous for each x in F. Conclude that $t \mapsto \varphi_t(a)$ is continuous for each a in A.

Exercise 3.8. Show that an element p in $M_2(\mathbb{C})$ is a one-dimensional projection if and only if

$$p = \begin{pmatrix} t & \omega\sqrt{t(1-t)} \\ \overline{\omega}\sqrt{t(1-t)} & 1-t \end{pmatrix},$$

for some t in $[0,1]$ and some complex number ω of modulus 1. Use this to show that the space $G_{2,1}$ of all one-dimensional projections in $M_2(\mathbb{C})$ is homeomorphic to the 2-sphere S^2.

Exercise 3.9. Let $\mathbf{C}$ be a category, see Paragraph 3.2.1. Two objects A and B in $\mathcal{O}(\mathbf{C})$ are said to be isomorphic, written $A \cong B$, if there exist morphisms φ in $\mathrm{Mor}(A, B)$ and ψ in $\mathrm{Mor}(B, A)$ such that $\psi \circ \varphi = \mathrm{id}_A$ and $\varphi \circ \psi = \mathrm{id}_B$.

 (i) Show that $\cong$ is an equivalence relation on $\mathcal{O}(\mathbf{C})$.

 (ii) Suppose that N and N' are zero objects in $\mathbf{C}$, see Paragraph 3.2.1. Show that $N \cong N'$.

 (iii) Suppose that $\mathbf{C}$ has a zero object N. Let $0_{N,A}$ and $0_{A,N}$ be the unique morphisms in $\mathrm{Mor}(A, N)$ and in $\mathrm{Mor}(N, A)$, respectively. For each pair of morphisms A and B in $\mathbf{C}$, set $0_{B,A} = 0_{B,N} \circ 0_{N,A}$ in $\mathrm{Mor}(A, B)$. Show that $0_{B,A}$ is independent of the choice of zero object N.

Exercise 3.10. Let R be a ring. For natural numbers m and n, the set of $m \times n$ matrices over R is denoted by $M_{m,n}(R)$ (or $M_n(R)$ in the case where $m = n$). An element e in R is *idempotent* if $e^2 = e$. We write $\mathcal{I}(R)$ for the set of idempotent elements in R, and we set

$$\mathcal{I}_n(R) = \mathcal{I}(M_n(R)), \qquad \mathcal{I}_\infty(R) = \bigcup_{n=1}^{\infty} \mathcal{I}_n(R).$$

Define the relation $\approx_0$ on $\mathcal{I}_\infty(R)$ as follows. Suppose that e belongs to $\mathcal{I}_n(R)$ and f to $\mathcal{I}_m(R)$. Then $e \approx_0 f$ if $e = ab$ and $f = ba$ for some elements a in $M_{n,m}(R)$ and b in $M_{m,n}(R)$.

(i) Suppose that $e \approx_0 f$, where e belongs to $\mathcal{I}_n(R)$ and f to $\mathcal{I}_m(R)$. Show that there are elements c in $M_{n,m}(R)$ and d in $M_{m,n}(R)$ such that $e = cd$, $f = dc$, $cdc = c$, and $dcd = d$. [Hint: Take $c = aba$ and $d = bab$, where a and b are as in the definition above.]

(ii) Show that $\approx_0$ is an equivalence relation on $\mathcal{I}_\infty(R)$.

(iii) Define an operation $\oplus$ on $\mathcal{I}_\infty(R)$ by

$$e \oplus f = \begin{pmatrix} e & 0 \\ 0 & f \end{pmatrix}, \qquad e, f \in \mathcal{I}_\infty(R).$$

Show that $e \oplus 0_n \approx_0 e$ for each e in $\mathcal{I}_\infty(R)$ and each natural number n, where 0_n is the zero element of $M_n(R)$.

(iv) Set $V(R) = \mathcal{I}_\infty(R)/\approx_0$. For each e in $\mathcal{I}_\infty(R)$, let $[e]_V$ denote the equivalence class of e in $V(R)$. Show that $\oplus$ induces an operation $+$ on $V(R)$ satisfying

$$[e]_V + [f]_V = [e \oplus f]_V, \qquad e, f \in \mathcal{I}_\infty(R).$$

(v) Show that $(V(R), +)$ is an Abelian semigroup.

In the case where R is a unital ring, $K_0(R)$ is defined to be the Grothendieck group of $(V(R), +)$.

In Exercise 3.11 it will be shown that this definition of K_0 agrees with the existing definition when A is a unital C^*-algebra.

Exercise 3.11. Let A be a C^*-algebra. The aim of this exercise is to show that the ring-theoretic definition of $K_0(A)$ given in Exercise 3.10 is essentially the same as the C^*-algebraic definition given in Definition 3.1.4. The terminology is the same as in Exercise 3.10.

(i) Show that, for every idempotent element e in A, there is a projection p in A with $e \approx_0 p$. [Hint: Observe that the element $h = 1 + (e - e^*)(e^* - e)$ in $\tilde{A}$ is invertible, and show that $eh = ee^*e = he$, $e^*h = e^*ee^* = he^*$. Set $p = ee^*h^{-1}$, and check that p is a projection in A satisfying $ep = p$ and $pe = e$.]

(ii) For projections p and q in A, show that $p \sim_0 q$ if and only if $p \approx_0 q$. [Hint: Take a and b in A with $p = ab$, $q = ba$, $a = aba$, and $b = bab$. Show that b^*b belongs to pAp and that $p \leqslant \|a\|^2 b^*b$. Deduce that $(b^*b)^{1/2}$

is invertible in pAp, take c in pAp such that $(b^*b)^{1/2}c = p = c(b^*b)^{1/2}$, and set $v = bc$. Show that $p = v^*v$ and $q = vv^*$; it may be helpful to verify that $qv = v$ first.]

(iii) Show that the map

$$[p]_{\mathcal{D}} \mapsto [p]_V, \quad \mathcal{D}(A) \to V(A),$$

is a well-defined semigroup isomorphism. [Hint: Observe that (i) and (ii) hold in any matrix algebra over A. To extend (ii) to projections in matrix algebras of different sizes, use Exercise 3.10 (iii).]

A somewhat related result can be found in Lemma 11.2.7.

Exercise 3.12. This exercise requires knowledge about von Neumann algebras. Let $\mathcal{M}$ be a von Neumann algebra factor of type II_1. Let τ be the unique normalized trace on $\mathcal{M}$. Show that

$$K_0(\tau): K_0(\mathcal{M}) \to \mathbb{R}$$

is an isomorphism, so that $K_0(\mathcal{M}) \cong \mathbb{R}$. [Hint: Use the following two facts about II_1-factors. Two projections in a II_1-factor are equivalent if (and only if) they have the same trace, and $\{\tau(p) : p \in \mathcal{P}(\mathcal{M})\} = [0, 1]$.]

Show that $K_0(\mathcal{M}) = 0$ if $\mathcal{M}$ is a factor of type II_∞ or of type III. (Factors of type I_∞ also have trivial K_0-group as shown in Example 3.3.3.)

Exercise 3.13. The purpose of this exercise is to verify the claim from Example 3.3.6 that the map $t \mapsto \varphi_t(f)$ is continuous for each f in $C(X)$, where X is a compact Hausdorff space, where $\alpha: [0, 1] \times X \to X$ is a continuous map, and where $\varphi_t(f)(x) = f(\alpha(t, x))$.

Let t_0 in $[0, 1]$ and $\varepsilon > 0$ be given. Set

$$W = \{(t, x) : |f(\alpha(t, x)) - f(\alpha(t_0, x))| < \varepsilon\}.$$

Show that there is $\delta > 0$ such that $(t_0 - \delta, t_0 + \delta) \times X$ is contained in W, and conclude that $\|\varphi_t(f) - \varphi_{t_0}(f)\| \leqslant \varepsilon$ for all t in $(t_0 - \delta, t_0 + \delta)$.

Chapter 4

The Functor K_0

We extend the functor K_0 considered in Chapter 3 to a functor from the category of all C^*-algebras, unital or not. The K_0-group of a non-unital C^*-algebra is defined as a relative K-group, and it is shown that this definition is coherent with the definition of K_0 for unital C^*-algebras.

We give a standard picture of K_0 (a concrete description of the elements in $K_0(A)$). This new standard picture is less pedestrian than the standard picture for K_0 for unital C^*-algebras stated in Proposition 3.1.7, reflecting the greater complexity of K_0 in the non-unital case.

It is shown that K_0 is a functor which is half exact, split exact, and stable. Examples are given to show that K_0 is not exact.

4.1 Definition and functoriality of K_0

Definition 4.1.1 (The K_0-group of non-unital C^*-algebras).   Let A be a non-unital C^*-algebra, and consider the associated split exact sequence

$$(4.1) \qquad 0 \longrightarrow A \overset{\iota}{\longrightarrow} \widetilde{A} \underset{\lambda}{\overset{\pi}{\rightleftarrows}} \mathbb{C} \longrightarrow 0$$

obtained by adjoining a unit to A (see Paragraph 1.1.6). Define $K_0(A)$ to be the kernel of the homomorphism $K_0(\pi) \colon K_0(\widetilde{A}) \to K_0(\mathbb{C})$.

Observe that $K_0(A)$ is an Abelian group, being a subgroup of $K_0(\widetilde{A})$.

For each p in $\mathcal{P}_\infty(A)$ consider the equivalence class $[p]_0$ in $K_0(\widetilde{A})$. Since $K_0(\pi)([p]_0) = [\pi(p)]_0 = 0$, it follows that $[p]_0$ belongs to $K_0(A)$. In this way we obtain a map $[\,\cdot\,]_0 \colon \mathcal{P}_\infty(A) \to K_0(A)$.

For each C^*-algebra A, unital or not, we have a short exact sequence

$$(4.2) \qquad 0 \longrightarrow K_0(A) \longrightarrow K_0(\widetilde{A}) \xrightarrow{\ K_0(\pi)\ } K_0(\mathbb{C}) \longrightarrow 0.$$

The map $K_0(A) \to K_0(\widetilde{A})$ is $K_0(\iota)$ when A is unital, and it is the inclusion map when A is non-unital. Use Lemma 3.2.8 to see that (4.2) is exact when A is unital.

When A is a unital C^*-algebra, $K_0(A)$ is isomorphic to its image in $K_0(\widetilde{A})$ under $K_0(\iota)$, and $K_0(\iota)$ maps $[p]_0$ in $K_0(A)$ to $[p]_0$ in $K_0(\widetilde{A})$ for p in $\mathcal{P}_\infty(A)$. The image of $K_0(\iota)$ is equal to the kernel of $K_0(\pi)$, so that the identity

$$K_0(A) = \mathrm{Ker}(K_0(\pi))$$

holds for both unital and non-unital C^*-algebras (with a slight abuse of notation).

4.1.2 The functor K_0. Let $\varphi: A \to B$ be a *-homomorphism, and let $\widetilde{\varphi}: \widetilde{A} \to \widetilde{B}$ be its unitization, see Paragraph 1.1.6. The commutative diagram

$$
\begin{array}{ccccc}
A & \xrightarrow{\iota_A} & \widetilde{A} & \xrightarrow{\pi_A} & \mathbb{C} \\
\varphi \downarrow & & \widetilde{\varphi} \downarrow & & \| \\
B & \xrightarrow{\iota_B} & \widetilde{B} & \xrightarrow{\pi_B} & \mathbb{C}
\end{array}
$$

and (4.2) induce by functoriality of K_0 for unital C^*-algebras the commutative diagram

$$(4.3) \qquad
\begin{array}{ccccc}
K_0(A) & \longrightarrow & K_0(\widetilde{A}) & \xrightarrow{K_0(\pi_A)} & K_0(\mathbb{C}) \\
K_0(\varphi) \downarrow & & K_0(\widetilde{\varphi}) \downarrow & & \| \\
K_0(B) & \longrightarrow & K_0(\widetilde{B}) & \xrightarrow[K_0(\pi_B)]{} & K_0(\mathbb{C}).
\end{array}
$$

There is one and only one group homomorphism $K_0(\varphi): K_0(A) \to K_0(B)$, indicated with a dashed arrow, that makes the diagram commutative, and $K_0(\varphi)$ is the restriction of $K_0(\widetilde{\varphi})$ to $K_0(A)$.

Suppose that both A and B are unital C^*-algebras. Then $K_0(\varphi)$ has already been defined in Paragraph 3.2.2. By functoriality of K_0 in the category of unital C^*-algebras (Proposition 3.2.4) the diagram (4.3), with $K_0(\varphi)$ from

Paragraph 3.2.2, is commutative. This shows that the present definition of $K_0(\varphi)$ extends Paragraph 3.2.2.

Notice that

(4.4) $K_0(\varphi)([p]_0) = [\varphi(p)]_0, \qquad p \in \mathcal{P}_\infty(A),$

when A is any C^*-algebra, unital or not.

Proposition 4.1.3 (i) and (ii) below say — in analogy with Proposition 3.2.4 — that K_0 is a functor from the category of C^*-algebras to the category of Abelian groups. Parts (iii) and (iv) say that K_0 maps zero objects to zero objects and zero morphisms to zero morphisms.

Proposition 4.1.3 (Functoriality of K_0).

(i) $K_0(\mathrm{id}_A) = \mathrm{id}_{K_0(A)}$ *for every C^*-algebra A.*

(ii) *If A, B, and C are C^*-algebras, and if $\varphi\colon A \to B$ and $\psi\colon B \to C$ are *-homomorphisms, then $K_0(\psi \circ \varphi) = K_0(\psi) \circ K_0(\varphi)$.*

(iii) $K_0(\{0\}) = \{0\}$.

(iv) $K_0(0_{B,A}) = 0_{K_0(B),K_0(A)}$ *for every pair of C^*-algebras A and B.*

Proof. Parts (i) and (ii) follow from Proposition 3.2.4 (i) and (ii) because $(\mathrm{id}_A)^\sim = \mathrm{id}_{\tilde{A}}$ and $(\psi \circ \varphi)^\sim = \tilde{\psi} \circ \tilde{\varphi}$.

(iii). We have $\widetilde{\{0\}} = \mathbb{C}$, and the sequence (4.1) in Definition 4.1.1, with $A = \{0\}$, becomes

$$0 \longrightarrow 0 \longrightarrow \mathbb{C} \stackrel{\pi}{\longrightarrow} \mathbb{C} \longrightarrow 0 \,,$$

where $\pi = \mathrm{id}_\mathbb{C}$. It follows that $K_0(\{0\}) = \mathrm{Ker}(K_0(\pi)) = \{0\}$.

(iv). This follows from (iii) as in the proof of Proposition 3.2.4.   □

Proposition 4.1.4 (Homotopy invariance of K_0). *Let A and B be C^*-algebras.*

(i) *If $\varphi, \psi\colon A \to B$ are homotopic *-homomorphisms, then $K_0(\varphi) = K_0(\psi)$.*

(ii) *If A and B are homotopy equivalent C^*-algebras, then $K_0(A)$ is isomorphic to $K_0(B)$. More specifically, if*

$$A \stackrel{\varphi}{\longrightarrow} B \stackrel{\psi}{\longrightarrow} A$$

is a homotopy, then $K_0(\varphi)\colon K_0(A) \to K_0(B)$ and $K_0(\psi)\colon K_0(B) \to K_0(A)$ are isomorphisms and $K_0(\varphi)^{-1} = K_0(\psi)$.

Proof. (i). If φ is homotopic to ψ, then $\widetilde{\varphi}$ is homotopic to $\widetilde{\psi}$, whence $K_0(\widetilde{\varphi}) = K_0(\widetilde{\psi})$ by Proposition 3.2.6. It follows that $K_0(\varphi)$ and $K_0(\psi)$ are equal, being restrictions of $K_0(\widetilde{\varphi})$ and $K_0(\widetilde{\psi})$ to $K_0(A)$.

(ii). This follows from (i) and from Proposition 4.1.3 (i) and (ii). $\qquad\square$

Example 4.1.5. The *cone*, CA, and the *suspension*, SA, of a C^*-algebra A are defined to be

$$CA = \{f \in C([0,1], A) : f(0) = 0\},$$
$$SA = \{f \in C([0,1], A) : f(0) = f(1) = 0\}.$$

We have a short exact sequence

$$0 \longrightarrow SA \overset{\iota}{\longrightarrow} CA \overset{\pi}{\longrightarrow} A \longrightarrow 0,$$

where ι is the inclusion mapping and $\pi(f) = f(1)$.

The cone CA is homotopy equivalent to the zero C^*-algebra. Indeed, set

$$\varphi_t \colon CA \to CA, \qquad \varphi_t(f)(s) = f(st), \quad f \in CA, \ s, t \in [0,1].$$

The map $t \mapsto \varphi_t(f)$ is continuous for each f in CA, $\varphi_0 = 0$ and $\varphi_1 = \mathrm{id}$. This proves that

$$CA \overset{0}{\longrightarrow} 0 \overset{0}{\longrightarrow} CA$$

is a homotopy. We infer from Propositions 4.1.3 and 4.1.4 that $K_0(CA) = 0$.

4.2   The standard picture of the group $K_0(A)$

Proposition 4.2.2 and Lemma 4.2.3 below contain the standard picture of K_0: a handy description of what elements in $K_0(A)$ look like. These results will be invoked whenever explicit calculations involving K_0-groups are performed.

4.2.1 The scalar mapping. Consider the split exact sequence

$$0 \longrightarrow A \overset{\iota}{\longrightarrow} \widetilde{A} \underset{\lambda}{\overset{\pi}{\rightleftarrows}} \mathbb{C} \longrightarrow 0$$

obtained by adjoining a unit to the C^*-algebra A. Define the *scalar mapping* s to be

$$s = \lambda \circ \pi \colon \widetilde{A} \to \widetilde{A},$$

i.e., $s(a + \alpha 1) = \alpha 1$ for all a in A and all α in $\mathbb{C}$. Notice that $\pi(s(x)) = \pi(x)$ and that $x - s(x)$ belongs to A for each x in $\widetilde{A}$. Let

$$s_n \colon M_n(\widetilde{A}) \to M_n(\widetilde{A})$$

be the *-homomorphism induced by s as in Section 1.3. The image of s_n is the subset $M_n(\mathbb{C})$ of $M_n(\widetilde{A})$ consisting of all matrices with scalar entries, and $x - s_n(x)$ belongs to $M_n(A)$ for all x in $M_n(\widetilde{A})$. We shall often omit the subscript and write s instead of s_n.

An element x in $\widetilde{A}$ or in $M_n(\widetilde{A})$ will be called a *scalar element* if $x = s(x)$. Equivalently, an element x in $M_n(\widetilde{A})$ is scalar if all its entries are scalar multiples of the unit $1_{\widetilde{A}}$.

The scalar mapping is natural in the sense that if A and B are C^*-algebras, and if $\varphi \colon A \to B$ is a *-homomorphism, then we get a commuting diagram

$$
\begin{array}{ccc}
\widetilde{A} & \xrightarrow{\ s\ } & \widetilde{A} \\
{\scriptstyle \widetilde{\varphi}}\downarrow & & \downarrow{\scriptstyle \widetilde{\varphi}} \\
\widetilde{B} & \xrightarrow{\ s\ } & \widetilde{B}.
\end{array}
$$

Proposition 4.2.2 (The standard picture of $K_0(A)$). *One has for each C^*-algebra A that*

(4.5) $$K_0(A) = \{[p]_0 - [s(p)]_0 : p \in \mathcal{P}_\infty(\widetilde{A})\}.$$

Moreover, the following hold.

(i) *For each pair of projections p, q in $\mathcal{P}_\infty(\widetilde{A})$, the following conditions are equivalent:*

  (a) $[p]_0 - [s(p)]_0 = [q]_0 - [s(q)]_0$,

  (b) *there exist natural numbers k and l such that $p \oplus 1_k \sim_0 q \oplus 1_l$ in $\mathcal{P}_\infty(\widetilde{A})$,*

  (c) *there exist scalar projections r_1 and r_2 such that $p \oplus r_1 \sim_0 q \oplus r_2$.*

(ii) *If p in $\mathcal{P}_\infty(\widetilde{A})$ satisfies $[p]_0 - [s(p)]_0 = 0$, then there is a natural number m with $p \oplus 1_m \sim s(p) \oplus 1_m$.*

(iii) *If $\varphi \colon A \to B$ is a *-homomorphism, then*

$$K_0(\varphi)([p]_0 - [s(p)]_0) = [\widetilde{\varphi}(p)]_0 - [s(\widetilde{\varphi}(p))]_0$$

for each p in $\mathcal{P}_\infty(\widetilde{A})$.

Proof. To prove that equation (4.5) holds, notice that

$$K_0(\pi)([p]_0 - [s(p)]_0) = [\pi(p)]_0 - [(\pi \circ s)(p)]_0 = 0$$

when p belongs to $\mathcal{P}_\infty(\widetilde{A})$ because $\pi = \pi \circ s$. This shows that $[p]_0 - [s(p)]_0$ belongs to $K_0(A)$ for all projections p in $\mathcal{P}_\infty(\widetilde{A})$.

Conversely, let g be an element in $K_0(A)$, and find n in $\mathbb{N}$ and projections e, f in $M_n(\widetilde{A})$ such that $g = [e]_0 - [f]_0$. Put

$$p = \begin{pmatrix} e & 0 \\ 0 & 1_n - f \end{pmatrix}, \qquad q = \begin{pmatrix} 0 & 0 \\ 0 & 1_n \end{pmatrix}.$$

Then p, q belong to $\mathcal{P}_{2n}(\widetilde{A})$, and

$$[p]_0 - [q]_0 = [e]_0 + [1_n - f]_0 - [1_n]_0 = [e]_0 - [f]_0 = g.$$

As $q = s(q)$ and $K_0(\pi)(g) = 0$, we deduce that

$$[s(p)]_0 - [q]_0 = [s(p)]_0 - [s(q)]_0 = K_0(s)(g) = (K_0(\lambda) \circ K_0(\pi))(g) = 0.$$

This shows that $g = [p]_0 - [s(p)]_0$.

(i). Let p, q in $\mathcal{P}_\infty(\widetilde{A})$ be given. Suppose that $[p]_0 - [s(p)]_0 = [q]_0 - [s(q)]_0$. Then $[p \oplus s(q)]_0 = [q \oplus s(p)]_0$, and hence $p \oplus s(q) \sim_s q \oplus s(p)$ in $\mathcal{P}_\infty(\widetilde{A})$ by Proposition 3.1.7. By the remark below Definition 3.1.6, there is a natural number n such that $p \oplus s(q) \oplus 1_n \sim_0 q \oplus s(p) \oplus 1_n$. This shows that (a) implies (c). To see that (c) implies (b) note that if r_1, r_2 are scalar projections in $\mathcal{P}_\infty(\widetilde{A})$ of dimension k and l, respectively, then $r_1 \sim_0 1_k$ and $r_2 \sim_0 1_l$ (see Exercise 2.9), and hence $p \oplus 1_k \sim_0 q \oplus 1_l$.

To see that (b) implies (a) note first that

$$[p \oplus 1_k]_0 - [s(p \oplus 1_k)]_0 = [p]_0 + [1_k]_0 - [s(p)]_0 - [1_k]_0 = [p]_0 - [s(p)]_0.$$

It therefore suffices to show that $[p]_0 - [s(p)]_0 = [q]_0 - [s(q)]_0$ when $p \sim_0 q$. Suppose accordingly that $p = v^*v$ and $q = vv^*$ for some partial isometry v in $M_{m,n}(\widetilde{A})$. Let $s(v)$ in $M_{m,n}(\mathbb{C})$, viewed as a subset of $M_{m,n}(\widetilde{A})$, be the matrix obtained by applying s to each entry of v — in analogy with the definition of s_n. Then $s(v)^*s(v) = s(p)$ and $s(v)s(v)^* = s(q)$, and so $s(p) \sim_0 s(q)$. Hence $[p]_0 = [q]_0$ and $[s(p)]_0 = [s(q)]_0$, and this proves that (a) holds.

(ii). If $[p]_0 - [s(p)]_0 = 0$, then $p \sim_s s(p)$ by Proposition 3.1.7 (v), and so there exists m in $\mathbb{N}$ such that $p \oplus 1_m \sim s(p) \oplus 1_m$, see the argument below Definition 3.1.6.

(iii). By definition, we have

$$K_0(\varphi)([p]_0 - [s(p)]_0) = K_0(\widetilde{\varphi})([p]_0 - [s(p)]_0)$$
$$= [\widetilde{\varphi}(p)]_0 - [\widetilde{\varphi}(s(p))]_0 = [\widetilde{\varphi}(p)]_0 - [s(\widetilde{\varphi}(p))]_0.$$

$\square$

We use the following, slightly technical, appendix to the standard picture to prove half exactness of K_0.

Lemma 4.2.3. *Let A and B be C^*-algebras, and let $\varphi\colon A \to B$ be a $*$-homomorphism. Suppose that g in $K_0(A)$ belongs to the kernel of $K_0(\varphi)$.*

(i) *There exist a natural number n, a projection p in $\mathcal{P}_n(\widetilde{A})$, and a unitary u in $M_n(\widetilde{B})$ such that $g = [p]_0 - [s(p)]_0$ and $u\widetilde{\varphi}(p)u^* = s(\widetilde{\varphi}(p))$.*

(ii) *If φ is surjective, then there is a projection p in $\mathcal{P}_\infty(\widetilde{A})$ satisfying $g = [p]_0 - [s(p)]_0$ and $\widetilde{\varphi}(p) = s(\widetilde{\varphi}(p))$.*

Proof. (i). It follows from the standard picture of K_0 (Proposition 4.2.2) that there exist a natural number k and a projection p_1 in $\mathcal{P}_k(\widetilde{A})$ with $g = [p_1]_0 - [s(p_1)]_0$ and $[\widetilde{\varphi}(p_1)]_0 - [s(\widetilde{\varphi}(p_1))]_0 = 0$. By Proposition 4.2.2 (ii), $\widetilde{\varphi}(p_1) \oplus 1_m \sim s(\widetilde{\varphi}(p_1)) \oplus 1_m$ for some m in $\mathbb{N}$. Put $p_2 = p_1 \oplus 1_m$. Then p_2 belongs to $\mathcal{P}_{k+m}(\widetilde{A})$, $g = [p_2]_0 - [s(p_2)]_0$, and

$$\widetilde{\varphi}(p_2) = \widetilde{\varphi}(p_1) \oplus 1_m \sim s(\widetilde{\varphi}(p_1)) \oplus 1_m = s(\widetilde{\varphi}(p_2)).$$

Put $n = 2(k + m)$. Let 0_{k+m} be the zero projection in $M_{k+m}(\widetilde{A})$, and set $p = p_2 \oplus 0_{k+m}$, so that p lies in $\mathcal{P}_n(\widetilde{A})$. It is clear that $g = [p]_0 - [s(p)]_0$, and it follows from Proposition 2.2.8 that $u\widetilde{\varphi}(p)u^* = s(\widetilde{\varphi}(p))$ for some unitary u in $M_n(\widetilde{B})$.

(ii). By (i) we can find n in $\mathbb{N}$, a projection p_1 in $M_n(\widetilde{A})$, and a unitary u in $M_n(\widetilde{B})$ with $g = [p_1]_0 - [s(p_1)]_0$ and $u\widetilde{\varphi}(p_1)u^* = s(\widetilde{\varphi}(p_1))$. Use Lemma 2.1.7 to find a unitary v in $M_{2n}(\widetilde{A})$ with $\widetilde{\varphi}(v) = \operatorname{diag}(u, u^*)$. Put $p = v \operatorname{diag}(p_1, 0_n) v^*$. Then p is a projection in $M_{2n}(\widetilde{A})$, and

$$\widetilde{\varphi}(p) = \begin{pmatrix} u & 0 \\ 0 & u^* \end{pmatrix} \begin{pmatrix} \widetilde{\varphi}(p_1) & 0 \\ 0 & 0 \end{pmatrix} \begin{pmatrix} u^* & 0 \\ 0 & u \end{pmatrix} = \begin{pmatrix} s(\widetilde{\varphi}(p_1)) & 0 \\ 0 & 0 \end{pmatrix} \in M_{2n}(\mathbb{C}).$$

It follows that $s(\widetilde{\varphi}(p)) = \widetilde{\varphi}(p)$. Finally, $g = [p]_0 - [s(p)]_0$ because $p \sim_0 p_1$, see Proposition 4.2.2 (i). $\square$

4.3   Half and split exactness and stability of K_0

We begin with a description of what happens when units are adjoined to a short exact sequence. The straightforward proof is left to the reader.

Lemma 4.3.1. *Let*

$$0 \longrightarrow I \xrightarrow{\ \varphi\ } A \xrightarrow{\ \psi\ } B \longrightarrow 0$$

be a short exact sequence of C^*-algebras, and let n be a natural number.

(i) *The mapping $\widetilde{\varphi}_n \colon M_n(\widetilde{I}) \to M_n(\widetilde{A})$ is injective.*

(ii) *An element a in $M_n(\widetilde{A})$ belongs to the image of $\widetilde{\varphi}_n$ if and only if $\widetilde{\psi}_n(a) = s_n(\widetilde{\psi}_n(a))$.*

Proposition 4.3.2 (Half exactness of K_0). *Every short exact sequence of C^*-algebras* $$0 \longrightarrow I \xrightarrow{\ \varphi\ } A \xrightarrow{\ \psi\ } B \longrightarrow 0$$

induces an exact sequence of Abelian groups

$$K_0(I) \xrightarrow{\ K_0(\varphi)\ } K_0(A) \xrightarrow{\ K_0(\psi)\ } K_0(B),$$

that is, $\mathrm{Im}(K_0(\varphi)) = \mathrm{Ker}(K_0(\psi))$.

Proof. Functoriality of K_0 yields that

$$K_0(\psi) \circ K_0(\varphi) = K_0(\psi \circ \varphi) = K_0(0) = 0.$$

Hence $\mathrm{Im}(K_0(\varphi))$ is contained in $\mathrm{Ker}(K_0(\psi))$.

Conversely, assume that g belongs to the kernel of $K_0(\psi)$. According to Lemma 4.2.3 (ii) we can find a natural number n and a projection p in $\mathcal{P}_n(\widetilde{A})$ with $g = [p]_0 - [s(p)]_0$ and $\widetilde{\psi}(p) = s(\widetilde{\psi}(p))$. By Lemma 4.3.1 (ii), there is an element e in $M_n(\widetilde{I})$ with $\widetilde{\varphi}(e) = p$. We also know from Lemma 4.3.1 that $\widetilde{\varphi}$ is injective. It follows that e must be a projection. Hence,

$$g = [\widetilde{\varphi}(e)]_0 - [s(\widetilde{\varphi}(e))]_0 = K_0(\varphi)([e]_0 - [s(e)]_0) \in \mathrm{Im}(K_0(\varphi)).$$

$\square$

Proposition 4.3.3 (Split exactness of K_0). *Every split exact sequence of C^*-algebras*

$$0 \longrightarrow I \overset{\varphi}{\longrightarrow} A \underset{\lambda}{\overset{\psi}{\rightleftarrows}} B \longrightarrow 0$$

induces a split exact sequence of Abelian groups

(4.6) $\qquad 0 \longrightarrow K_0(I) \overset{K_0(\varphi)}{\longrightarrow} K_0(A) \underset{K_0(\lambda)}{\overset{K_0(\psi)}{\rightleftarrows}} K_0(B) \longrightarrow 0.$

Proof. The sequence (4.6) is exact at $K_0(A)$ by Proposition 4.3.2. From functoriality of K_0 we get

$$\mathrm{id}_{K_0(B)} = K_0(\mathrm{id}_B) = K_0(\psi) \circ K_0(\lambda).$$

Hence (4.6) is exact at $K_0(B)$. We show below that $K_0(\varphi)$ is injective, and this will imply exactness of (4.6) at $K_0(I)$.

Let g in the kernel of $K_0(\varphi)$ be given. By the appendix to the standard picture of K_0 (Lemma 4.2.3 (i)) we can find a natural number n, a projection p in $\mathcal{P}_n(\tilde{I})$, and a unitary u in $M_n(\tilde{A})$ with $g = [p]_0 - [s(p)]_0$ and $u\tilde{\varphi}(p)u^* = s(\tilde{\varphi}(p))$. Put $v = (\tilde{\lambda} \circ \tilde{\psi})(u^*)u$. Then v is a unitary element in $M_n(\tilde{A})$ and $\tilde{\psi}(v) = 1$. By Lemma 4.3.1 (ii) there is an element w in $M_n(\tilde{I})$ with $\tilde{\varphi}(w) = v$. Since $\tilde{\varphi}$ is injective, w must be unitary. From the calculation

$$\tilde{\varphi}(wpw^*) = v\tilde{\varphi}(p)v^* = (\tilde{\lambda} \circ \tilde{\psi})(u^*)s(\tilde{\varphi}(p))(\tilde{\lambda} \circ \tilde{\psi})(u)$$
$$= (\tilde{\lambda} \circ \tilde{\psi})(u^*s(\tilde{\varphi}(p))u) = (\tilde{\lambda} \circ \tilde{\psi})(\tilde{\varphi}(p)) = s(\tilde{\varphi}(p)) = \tilde{\varphi}(s(p))$$

and by injectivity of $\tilde{\varphi}$ we conclude that $wpw^* = s(p)$. This shows that $p \sim_u s(p)$ in $M_n(\tilde{I})$, and hence that $g = 0$. $\qquad\square$

One could also prove split exactness by using the long exact sequence in K-theory developed in Chapter 9. (See Exercise 9.1.)

Proposition 4.3.4 (Direct sums). *For every pair of C^*-algebras A and B, we have*

$$K_0(A \oplus B) \cong K_0(A) \oplus K_0(B).$$

More specifically, if $\iota_A \colon A \to A \oplus B$ and $\iota_B \colon B \to A \oplus B$ are the canonical inclusion mappings, then the map

$$K_0(\iota_A) \oplus K_0(\iota_B) \colon K_0(A) \oplus K_0(B) \to K_0(A \oplus B),$$

which maps (g, h) in $K_0(A) \oplus K_0(B)$ to $K_0(\iota_A)(g) + K_0(\iota_B)(h)$, is an isomorphism.

Proof. Consider the diagram

$$
\begin{array}{ccccccccc}
0 & \longrightarrow & K_0(A) & \xrightarrow{\ \alpha\ } & K_0(A) \oplus K_0(B) & \xrightarrow{\ \beta\ } & K_0(B) & \longrightarrow & 0 \\
  &   & \| &   & \Big\downarrow{\scriptstyle K_0(\iota_A) \oplus K_0(\iota_B)} & & \| & & \\
0 & \longrightarrow & K_0(A) & \xrightarrow[K_0(\iota_A)]{} & K_0(A \oplus B) & \xrightarrow[K_0(\pi_B)]{} & K_0(B) & \longrightarrow & 0
\end{array}
$$

where $\alpha(g) = (g, 0)$, $\beta(g, h) = h$, and where $\pi_B(a, b) = b$. The rows in the diagram are exact (the lower one by Proposition 4.3.3), and the diagram is commutative because $\pi_B \circ \iota_A = 0$ and $\pi_B \circ \iota_B = \mathrm{id}_B$. An easy diagram chase (or the five lemma) shows that $K_0(\iota_A) \oplus K_0(\iota_B)$ is an isomorphism. $\quad\square$

Example 4.3.5. From split exactness of K_0 (Proposition 4.3.3) and Example 3.3.2 we get for every C^*-algebra A that

$$K_0(\widetilde{A}) \cong K_0(A) \oplus \mathbb{Z}.$$

The next two examples show that the functor K_0 is not exact, see Paragraph 3.3.8.

Example 4.3.6. Consider the sequence

$$0 \longrightarrow C_0((0, 1)) \xrightarrow{\ \iota\ } C([0, 1]) \xrightarrow{\ \psi\ } \mathbb{C} \oplus \mathbb{C} \longrightarrow 0$$

from Exercise 1.2. From Proposition 4.3.4 and Example 3.3.2, $K_0(\mathbb{C} \oplus \mathbb{C}) \cong \mathbb{Z}^2$, and from Example 3.3.6 and Exercise 3.3, $K_0(C([0, 1])) \cong \mathbb{Z}$. Therefore $K_0(\psi)$ cannot be surjective.

Example 4.3.7. Let H be a separable infinite dimensional Hilbert space, and let $\mathcal{K}$ be the ideal of compact operators in $B(H)$. The quotient $B(H)/\mathcal{K}$ is called the *Calkin algebra*, and is sometimes denoted by $\mathcal{Q}(H)$. We have a short exact sequence

$$0 \longrightarrow \mathcal{K} \xrightarrow{\ \iota\ } B(H) \xrightarrow{\ \pi\ } \mathcal{Q}(H) \longrightarrow 0.$$

From Example 3.3.3 we have $K_0(B(H)) = 0$. It will be shown in Corollary 6.4.2 that $K_0(\mathcal{K})$ is isomorphic to $\mathbb{Z}$. Therefore $K_0(\iota)$ cannot be injective.

Proposition 4.3.8* (**Stability of K_0**). *Let A be a C^*-algebra and let n be a natural number. Then $K_0(A)$ is isomorphic to $K_0(M_n(A))$.*
  More specifically, the $*$-homomorphism

$$\lambda_{n,A} \colon A \to M_n(A), \qquad a \mapsto \begin{pmatrix} a & 0 \\ 0 & 0 \end{pmatrix},$$

induces an isomorphism $K_0(\lambda_{n,A}) \colon K_0(A) \to K_0(M_n(A))$.

We shall rephrase stability of K_0 in terms of stabilized C^*-algebras in Proposition 6.4.1.

Proof of Proposition. We show first that the non-unital case can be derived from the unital case. For this purpose let A be a (non-unital) C^*-algebra. Then

$$
\begin{array}{ccccccccc}
0 & \longrightarrow & A & \longrightarrow & \widetilde{A} & \overset{\pi}{\longrightarrow} & \mathbb{C} & \longrightarrow & 0 \\
& & \downarrow{\lambda_{n,A}} & & \downarrow{\lambda_{n,\widetilde{A}}} & & \downarrow{\lambda_{n,\mathbb{C}}} & & \\
0 & \longrightarrow & M_n(A) & \longrightarrow & M_n(\widetilde{A}) & \overset{\pi_n}{\longrightarrow} & M_n(\mathbb{C}) & \longrightarrow & 0
\end{array}
$$

is a commutative diagram with split exact rows. It follows that

$$
\begin{array}{ccccccccc}
0 & \longrightarrow & K_0(A) & \longrightarrow & K_0(\widetilde{A}) & \longrightarrow & K_0(\mathbb{C}) & \longrightarrow & 0 \\
& & \downarrow{K_0(\lambda_{n,A})} & & \downarrow{K_0(\lambda_{n,\widetilde{A}})} & & \downarrow{K_0(\lambda_{n,\mathbb{C}})} & & \\
0 & \longrightarrow & K_0(M_n(A)) & \longrightarrow & K_0(M_n(\widetilde{A})) & \longrightarrow & K_0(M_n(\mathbb{C})) & \longrightarrow & 0
\end{array}
$$

is a commutative diagram with (split) exact rows. A diagram chase (or the five lemma) now shows that $K_0(\lambda_{n,A})$ is an isomorphism if $K_0(\lambda_{n,\widetilde{A}})$ and $K_0(\lambda_{n,\mathbb{C}})$ are isomorphisms. We therefore need only prove the proposition when A is unital.

    The short version of the rest of the proof is to remark that the two sets of projections $\mathcal{P}_\infty(A)$ and $\mathcal{P}_\infty(M_n(A))$ for all practical purposes are identical! We formalize this idea by constructing an inverse to the map $K_0(\lambda_{n,A})$ as follows. For each natural number k let $\gamma_{n,k} \colon M_k(M_n(A)) \to M_{kn}(A)$ be the $*$-isomorphism given by viewing each element of $M_k(M_n(A))$ as one big matrix (in $M_{kn}(A)$) — in other words, $\gamma_{n,k}$ erases certain parentheses. Define $\gamma_n \colon \mathcal{P}_\infty(M_n(A)) \to K_0(A)$ by $\gamma_n(p) = [\gamma_{n,k}(p)]_0$ for p in $\mathcal{P}_k(M_n(A))$. Apply the universal property of K_0 (Proposition 3.1.8) to γ_n (and verify conditions (i), (ii), and (iii) in Proposition 3.1.8) to obtain a group homomorphism $\alpha\colon K_0(M_n(A)) \to K_0(A)$ satisfying $\alpha([p]_0) = [\gamma_{n,k}(p)]_0$ for p in $\mathcal{P}_k(M_n(A))$.

We claim that α is the inverse to $K_0(\lambda_{n,A})$, and to prove this claim it suffices to show that

$$(4.7) \qquad \begin{aligned} (\lambda_{n,A})_{kn}\big(\gamma_{n,k}(p)\big) &\sim_0 p \quad \text{in} \quad \mathcal{P}_\infty(M_n(A)), \quad p \in \mathcal{P}_k(M_n(A)), \\ \gamma_{n,k}\big((\lambda_{n,A})_k(p)\big) &\sim_0 p \quad \text{in} \quad \mathcal{P}_\infty(A), \qquad p \in \mathcal{P}_k(A), \end{aligned}$$

where $(\lambda_{n,A})_m$ is the *-homomorphism $M_m(A) \to M_m(M_n(A))$ induced by $\lambda_{n,A}$. We prove the second claim in (4.7); the proof of the first claim in (4.7) is similar.

Let $\{e_1, e_2, \ldots, e_{kn}\}$ be the standard basis for $\mathbb{C}^{kn}$, and let u be a permutation unitary in $M_{kn}(\mathbb{C}) \subseteq M_{kn}(A)$ that fulfills

$$ue_i = e_{n(i-1)+1}, \qquad i = 1, 2, \ldots, k.$$

Then

$$p \sim_0 p \oplus 0_{(n-1)k} = u^* \gamma_{n,k}\big((\lambda_{n,A})_k(p)\big) u$$

for all projections p in $\mathcal{P}_k(A)$. $\square$

4.4 Exercises

Exercise 4.1. Show that $K_0(A)$ is a countable Abelian group if A is a separable C^*-algebra. [Hint: Use Exercise 3.1.]

Exercise 4.2. Let X be a locally compact Hausdorff space, and suppose that $X = X_1 \cup X_2$, where X_1, X_2 are all disjoint open and closed subsets of X. Show that $C_0(X)$ is isomorphic to $C_0(X_1) \oplus C_0(X_2)$.

It will be shown later that dim$\colon K_0(C(\mathbb{T})) \to \mathbb{Z}$ is an isomorphism. Use this to show that $K_0(C_0(\mathbb{R})) = 0$. Calculate $K_0(C_0(U))$, where

$$U = [-3, 0] \cup (1, 3) \cup (5, 6) \cup (8, 11] \cup \{15\}.$$

Exercise 4.3. Let A be a unital C^*-algebra, and let s be an isometry in A. Show that the map $\mu\colon A \to A$ given by $\mu(a) = sas^*$ is an endomorphism on A (i.e., a *-homomorphism from A into itself). Show that $K_0(\mu) = \mathrm{id}$. [Hint: Show that $\mu_n\colon M_n(A) \to M_n(A)$ is given by $\mu_n(a) = s_n a s_n^*$, where $s_n = \mathrm{diag}(s, s, \ldots, s)$.]

Exercise 4.4 (On the standard picture of $K_0(A)$). Let A be a C^*-algebra, unital or not. Show that each element in $K_0(A)$ is of the form

$$[p]_0 - \left[\begin{pmatrix} 1_n & 0_n \\ 0_n & 0_n \end{pmatrix} \right]_0$$

for some projection p in $M_{2n}(\widetilde{A})$ satisfying

(4.8) $$p - \begin{pmatrix} 1_n & 0_n \\ 0_n & 0_n \end{pmatrix} \in M_{2n}(A).$$

Show that an element p in $M_{2n}(\widetilde{A})$ satisfies (4.8) if and only if $s(p) = \mathrm{diag}(1_n, 0_n)$.

Exercise 4.5 (The Cuntz algebras $\mathcal{O}_n$). Let $n \geqslant 2$ be an integer, and let H be a separable, infinite dimensional Hilbert space.

(i) Show that there are elements $s_1, s_2, \ldots, s_n$ in $B(H)$ satisfying

$$s_1^* s_1 = s_2^* s_2 = \cdots = s_n^* s_n = 1 = s_1 s_1^* + s_2 s_2^* + \cdots + s_n s_n^*.$$

[Hint: Decompose H as $H = H_1 \oplus H_2 \oplus \cdots \oplus H_n$ for suitable subspaces $H_1, H_2, \ldots, H_n$ of H.]

Set $\mathcal{O}_n = C^*(s_1, s_2, \ldots, s_n)$, i.e., $\mathcal{O}_n$ is the sub-C^*-algebra of $B(H)$ generated by the isometries $s_1, s_2, \ldots, s_n$. The C^*-algebra $\mathcal{O}_n$ is called the *Cuntz algebra* (of order n). It was introduced by J. Cuntz in [12], where he proved that $\mathcal{O}_n$ is simple and has the following universal property: if A is a unital C^*-algebra containing elements $t_1, t_2, \ldots, t_n$ satisfying

$$t_j^* t_j = 1 = t_1 t_1^* + t_2 t_2^* + \cdots + t_n t_n^*,$$

then there is precisely one *-homomorphism $\varphi \colon \mathcal{O}_n \to A$ satisfying $\varphi(s_j) = t_j$.

You may use these non-trivial facts when answering the following questions.

(ii) Let u be a unitary element in $\mathcal{O}_n$. Show that there is precisely one unital endomorphism φ_u on $\mathcal{O}_n$ such that $\varphi_u(s_j) = u s_j$ for $j = 1, 2, \ldots, n$, and show that

(4.9) $$u = \sum_{j=1}^n \varphi_u(s_j) s_j^*.$$

(iii) Let φ be a unital endomorphism on $\mathcal{O}_n$. Show that $\varphi = \varphi_u$ for some unitary u in $\mathcal{O}_n$. [Hint: Look at (4.9).]

Let $\lambda \colon \mathcal{O}_n \to \mathcal{O}_n$ be given by

$$\lambda(x) = \sum_{j=1}^{n} s_j x s_j^*.$$

(iv) Show that λ is an endomorphism on $\mathcal{O}_n$, and show that $K_0(\lambda)(g) = ng$ for all g in $K_0(\mathcal{O}_n)$. [Hint: See Exercise 4.3.]

(v) Find the unitary element u in $\mathcal{O}_n$ for which $\lambda = \varphi_u$, see question (iii). Show that $u = u^*$ and conclude that $u \sim_h 1$ in $\mathcal{U}(\mathcal{O}_n)$. Use this to show that $\lambda \sim_h \mathrm{id}$, and hence that $K_0(\lambda)(g) = g$ for all g in $K_0(\mathcal{O}_n)$. [Hint: Use Exercise 3.7.]

(vi) Show that $K_0(\mathcal{O}_2) = 0$, and show that $(n-1)g = 0$ for all g in $K_0(\mathcal{O}_n)$.

It is shown in [13] that $K_0(\mathcal{O}_n) \cong \mathbb{Z}/(n-1)\mathbb{Z}$, and, more explicitly, that

$$K_0(\mathcal{O}_n) = \{0, [1]_0, 2[1]_0, \ldots, (n-2)[1]_0\}.$$

Exercise 4.6 (Properly infinite projections). A non-zero projection p in a C^*-algebra A is said to be *properly infinite* if there are mutually orthogonal projections e, f in A such that $e \leqslant p$, $f \leqslant p$, and $p \sim e \sim f$. A unital C^*-algebra A is said to be *properly infinite* if 1_A is a properly infinite projection.

Assume in the following that A is a properly infinite unital C^*-algebra.

(i) Show that A contains isometries s_1, s_2 with $s_1 s_1^* \perp s_2 s_2^*$.

(ii) Show that A contains a sequence of isometries $\{t_i\}_{i=1}^{\infty}$ with $t_j t_j^* \perp t_i t_i^*$ when $i \neq j$. [Hint: Look at $s_1, s_2 s_1, s_2^2 s_1, \ldots$]

(iii) For each n in $\mathbb{N}$, let v_n in $M_{1,n}(A)$ be the row matrix with entries $t_1, t_2, \ldots, t_n$, where $\{t_i\}$ is as in (ii). Show that $v_n^* v_n = 1_n$, the unit in $M_n(A)$.

(iv) Let p in $\mathcal{P}_n(A)$ be given, and let v_n be as in (iii). Show that $v_n p v_n^*$ is a projection in A and that $p \sim_0 v_n p v_n^*$.

(v) Let p, q be projections in A. Put

$$r = t_1 p t_1^* + t_2(1 - q)t_2^* + t_3(1 - t_1 t_1^* - t_2 t_2^*)t_3^*.$$

Show that r is a projection in A and that $[r]_0 = [p]_0 - [q]_0$.

(vi) Use (iv) and (v) to show that

$$K_0(A) = \{[p]_0 : p \,\epsilon\, \mathcal{P}(A), \, p \neq 0\}.$$

(vii) Show that the Cuntz algebra $\mathcal{O}_n$ (from Exercise 4.5) is properly infinite. Let H be a Hilbert space. Show that $B(H)$ is properly infinite if and only if H is infinite dimensional.

Exercise 4.7 (The relation $\precsim$ on $\mathcal{P}_\infty(A)$). Let A be a C^*-algebra and define a relation $\precsim$ on $\mathcal{P}_\infty(A)$ as follows. For p in $\mathcal{P}_n(A)$ and q in $\mathcal{P}_m(A)$, $p \precsim q$ if there is a projection q_0 in $\mathcal{P}_m(A)$ such that $p \sim_0 q_0 \leqslant q$.

(i) Show that $p \precsim q$ if and only if $q \sim_0 p \oplus p_0$ for some projection p_0 in $\mathcal{P}_\infty(A)$.

(ii) Show that $\precsim$ is transitive.

(iii) Let p_1, p_2, q_1, q_2 in $\mathcal{P}_\infty(A)$ be such that $p_j \precsim q_j$ for $j = 1, 2$. Show that $p_1 \oplus p_2 \precsim q_1 \oplus q_2$.

(iv) Show that a non-zero projection p in $\mathcal{P}_\infty(A)$ is properly infinite (see Exercise 4.6) if and only if $p \oplus p \precsim p$.

(v) Let p, q be projections in A, and suppose that $p \precsim q \precsim p$ and that p is properly infinite. Show that q is properly infinite.

Note. It is not true that $p \precsim q \precsim p$ implies $p \sim_0 q$. For example in the Cuntz algebra $\mathcal{O}_n$ (see Exercise 4.5) one has $p \precsim q \precsim p$ for each pair of non-zero projections p and q, but not all projections in $\mathcal{O}_n$ are equivalent if $n \geqslant 3$. (See [13] for further details.)

Exercise 4.8 (Full projections). Let A be a unital C^*-algebra. An element a in A is said to be *full* if it is not contained in any proper, closed, two-sided ideal in A.

(i) Let a be a full element in A. Show that there exist n in N and $x_1, x_2, \ldots, x_n$, $y_1, y_2, \ldots, y_n$ in A such that $1_A = \sum_{j=1}^{n} x_j a y_j$. [Hint: Show that the closure of the set

$$I_{\text{alg}}(a) = \left\{ \sum_{j=1}^{n} x_j a y_j : n \in \mathbb{N}, \ x_j, y_j \in A \right\}$$

is a closed, two-sided ideal in A, and conclude that $I_{\text{alg}}(a)$ is dense in A. Use this to show that $I_{\text{alg}}(a)$ contains an invertible element.]

(ii) Let a be an element in A^+ and let x, y be in A. Show that

$$xay + y^* a x^* \leqslant x a x^* + y^* a y.$$

(iii) Let a be an element in A^+ with $a \geqslant 1_A$. Show that there exists r in A with $1_A = rar^*$.

(iv) Use (i), (ii), and (iii) to show that if a is a full, positive element in A, then for some n in N there exist $x_1, x_2, \ldots, x_n$ in A such that

$$1_A = \sum_{j=1}^{n} x_j a x_j^*.$$

(v) Let a be a full, positive element in A, and let q be an arbitrary projection in A. Use (iii) to show that there exist n in N and $x_1, x_2, \ldots, x_n$ in A such that

$$q = \sum_{j=1}^{n} x_j a x_j^*.$$

(vi) Show that if p, q are projections in A where p is full, then there exists a natural number n such that $q \precsim p \oplus p \oplus \cdots \oplus p$ (with n summands). [Hint: Use (v) to find v in $M_{1,n}(A)$ such that $vv^* = q$ and $v^* v \leqslant p \oplus p \oplus \cdots \oplus p$.]

(vii) Show that A is properly infinite if it contains a properly infinite, full projection. [Hint: Use (vi) and Exercise 4.7.]

Exercise 4.9 (Properly infinite, full projections).   Let A be a unital C^*-algebra. A projection p in $\mathcal{P}_n(A) \subseteq \mathcal{P}_\infty(A)$ is said to be full if it is a full projection in $M_n(A)$ (see Exercise 4.8).

(i) Show that $q \precsim p$ for every projection q in $\mathcal{P}_\infty(A)$ if p is a properly infinite, full projection in $\mathcal{P}_\infty(A)$ (see Exercise 4.6). [Hint: We have p belongs to $\mathcal{P}_n(A)$ for some natural number n. First use Exercise 4.8 (vi) and Exercise 4.7 (iv) to show that $1_n \precsim p$, and then use Exercise 4.7 (iv) again.]

(ii) Suppose that p, q are projections in $\mathcal{P}_\infty(A)$ and that $[p]_0 = [q]_0$. Show that $p \oplus e \sim_0 q \oplus e$ for each properly infinite, full projection e in $\mathcal{P}_\infty(A)$.

(iii) Let p, q be properly infinite, full projections in $\mathcal{P}_\infty(A)$. Show that $[p]_0 = [q]_0$ if and only if $p \sim_0 q$. [Hint: Use (i) and Exercise 4.7 (i) to find a projection q_0 in $\mathcal{P}_\infty(A)$ such that $q \sim_0 q_0 \oplus p$, observe that $p \sim_0 0 \oplus p$, and use (ii).]

Exercise 4.10. Let A be a C^*-algebra, and let $\mathrm{Aut}(A)$ denote the group of *-automorphisms on A.

(i) Let u be a unitary in $\mathcal{U}(\widetilde{A})$. Show that the map

$$\mathrm{Ad}\, u \colon A \to A, \qquad a \mapsto uau^*,$$

defines an automorphism on A.

(ii) Show that the map $\mathcal{U}(\widetilde{A}) \to \mathrm{Aut}(A)$, $u \mapsto \mathrm{Ad}\, u$, is a group homomorphism.

Put $\mathrm{Inn}(A) = \{\mathrm{Ad}\, u : u \in \mathcal{U}(\widetilde{A})\}$. The elements in $\mathrm{Inn}(A)$ are called *inner automorphisms*.

(iii) Show that $\mathrm{Inn}(A)$ is a normal subgroup of $\mathrm{Aut}(A)$.

(iv) Show that if A is unital, then every inner automorphism has the form $\mathrm{Ad}\, u$ for some unitary u that belongs to A (rather than to $\widetilde{A}$).

(v) Show that $K_0(\alpha) = \mathrm{id}$ for every inner automorphism α.

An automorphism α on A is called *approximately inner* if, for every finite subset F of A and for every $\varepsilon > 0$, there is an inner automorphism β on A such that $\|\alpha(a) - \beta(a)\| < \varepsilon$ for all a in F. One can check that if A is separable and if α is an automorphism on A, then α is approximately inner if and only if there exists a sequence $\{\alpha_n\}_{n=1}^\infty$ in $\mathrm{Inn}(A)$ converging to α point-wise, i.e., $\alpha_n(a) \to \alpha(a)$ for all a in A.

The set of approximately inner automorphisms is denoted by $\overline{\mathrm{Inn}}(A)$.

(vi) Show that $\overline{\mathrm{Inn}}(A)$ is a normal subgroup of $\mathrm{Aut}(A)$.

(vii) Show that $K_0(\alpha) = \mathrm{id}$ for all approximately inner automorphisms α.

(viii) Show that $K_0(\alpha)$ is an automorphism on $K_0(A)$ for all automorphisms α on A.

(ix) Show that the map $\mathrm{Aut}(A) \to \mathrm{Aut}(K_0(A))$, $\alpha \mapsto K_0(\alpha)$, is a group homomorphism.

(x) Give an example of a C^*-algebra A and an automorphism α on A such that $K_0(\alpha) \neq \mathrm{id}$. [Hint: Try $A = \mathbb{C} \oplus \mathbb{C}$.]

Exercise 4.11. Let A be a unital C^*-algebra, and let a be a positive element in A with $\|a\| \leqslant 1$. Notice that $a(1 - a)$ is positive, and show that

$$p = \begin{pmatrix} a & (a - a^2)^{1/2} \\ (a - a^2)^{1/2} & 1 - a \end{pmatrix}$$

is a projection in $M_2(A)$. Show that

$$p \sim \begin{pmatrix} 1 & 0 \\ 0 & 0 \end{pmatrix}.$$

Let B be another unital C^*-algebra, and let $\varphi \colon A \to B$ be a unit preserving, surjective *-homomorphism. Let q be a projection in B. Show that there is a positive element a in A with $\|a\| \leqslant 1$ such that $\varphi(a) = q$. [Hint: See Paragraph 2.2.10.] Let p be as above. Show that

$$\varphi(p) = \begin{pmatrix} q & 0 \\ 0 & 1 - q \end{pmatrix}.$$

Chapter 5

The Ordered Abelian Group $K_0(A)$

An extra structure is added to the Abelian group $K_0(A)$ of a C^*-algebra A by specifying a certain subset of it, called $K_0(A)^+$. The set $K_0(A)^+$ consists of all elements in $K_0(A)$ of the form $[p]_0$, where p is a projection in $\mathcal{P}_\infty(A)$. When A is a unital, stably finite C^*-algebra, then $(K_0(A), K_0(A)^+)$ has the pleasant structure of an ordered Abelian group. We shall for this purpose also discuss finiteness properties of C^*-algebras and of projections.

5.1   The ordered K_0-group of stably finite C^*-algebras

An element a in a unital C^*-algebra A is called *left-invertible* if there exists an element b in A such that $ba = 1$, and a is called *right-invertible* if $ab = 1$ for some b in A. If $b_1 a = ab_2 = 1$, then $b_1 = b_1 ab_2 = b_2$, and this shows that a is invertible if and only if a is both left- and right-invertible. Moreover, a is left-invertible if and only if $a^* a$ is invertible, and, similarly, a is right-invertible if and only if aa^* is invertible. (See Exercise 5.1).

Definition 5.1.1. A projection p in a C^*-algebra A is said to be *infinite* if it is equivalent to a proper subprojection of itself, i.e., if there is a projection q in A such that $p \sim q < p$. If p is not infinite, then p is said to be *finite*.

A unital C^*-algebra A is said to be *finite* if its unit 1_A is a finite projection. Otherwise A is called *infinite*. If $M_n(A)$ is finite for all positive integers n, then A is *stably finite*.

If A is a C^*-algebra without a unit, then A is called finite/stably finite/infinite if its unitization $\widetilde{A}$ is finite/stably finite/infinite.

A projection p in a C^*-algebra A is finite if and only if the C^*-algebra pAp is finite. Recall that an element s in a unital C^*-algebra is an isometry if $s^*s = 1$.

Lemma 5.1.2. *The following five conditions are equivalent for any unital C^*-algebra A.*

(i) *A is finite.*

(ii) *All isometries in A are unitary.*

(iii) *All projections in A are finite.*

(iv) *Every left-invertible element in A is invertible.*

(v) *Every right-invertible element in A is invertible.*

Proof. (i) $\Rightarrow$ (ii). If s in A is an isometry, then $1 = s^*s \sim ss^* \leqslant 1$. Hence $ss^* = 1$ and s is unitary.

(ii) $\Rightarrow$ (i). If $1 \sim p$ for some projection p in A, then $s^*s = 1$ and $ss^* = p$ for some s in A. Now, s is an isometry and therefore unitary. Hence $p = ss^* = 1$, and this shows that 1 is a finite projection, and hence that A is finite.

(ii) $\Rightarrow$ (iii). Suppose that p, q are projections in A satisfying $p \sim q$ and $q \leqslant p$. Let v be a partial isometry in A which implements the equivalence between p and q: $v^*v = p$ and $vv^* = q$. Put $s = v + (1 - p)$. From (2.3) and the assumption that $pq = qp = q$ we find that $v^*(1 - p) = 0 = (1 - p)v$. We conclude that $s^*s = 1$ and $ss^* = 1 - (p - q)$. Hence, if (ii) holds, then $p = q$, proving (iii).

(iii) $\Rightarrow$ (i). If (iii) holds, then the unit of A is a finite projection.

(iv) $\Longleftrightarrow$ (v). If a in A is left-invertible and not right-invertible, then a^* is right-invertible and not left-invertible.

(iv) $\Rightarrow$ (ii). This is trivial.

(ii) $\Rightarrow$ (iv). Suppose that (ii) holds, and let a be a left-invertible element in A. Find b in A with $ba = 1$. Recall that if h, h_1, h_2 are self-adjoint elements in a unital C^*-algebra, then $h \leqslant \|h\| \cdot 1$; and if $h_1 \leqslant h_2$, then $x^*h_1x \leqslant x^*h_2x$ for each element x in the C^*-algebra. Use these facts to obtain the inequality

$$1 = (ba)^*(ba) = a^*b^*ba \leqslant \|b\|^2 a^*a.$$

This shows that $a^*a - \|b\|^{-2}\cdot 1 \geqslant 0$, and the spectrum of a^*a is therefore contained in the interval $[\|b\|^{-2}, \infty)$, which in particular implies that a^*a is invertible. Put $s = a(a^*a)^{-1/2}$, and observe that

$$s^*s = (a^*a)^{-1/2}a^*a(a^*a)^{-1/2} = 1.$$

Then s is invertible by (ii), and hence $a = s(a^*a)^{1/2}$ is invertible.   □

A finite C^*-algebra — unital or not — will always satisfy Lemma 5.1.2 (iii). The converse does not hold. A non-unital C^*-algebra can be infinite and satisfy Lemma 5.1.2 (iii) (see Example 9.4.4). A finite C^*-algebra need not be stably finite (Exercise 9.7).

Definition 5.1.3. A pair (G, G^+) is called an *ordered Abelian group* if G is an Abelian group, G^+ is a subset of G, and

(5.1)   (i) $G^+ + G^+ \subseteq G^+$,   (ii) $G^+ \cap (-G^+) = \{0\}$,   (iii) $G^+ - G^+ = G$.

Define a relation $\leqslant$ on G by $x \leqslant y$ if $y - x$ belongs to G^+.

Conditions (5.1) (i) and (ii) imply that $(G, \leqslant)$ is a (partially) ordered set. Conversely, if $\leqslant$ is an order relation on an Abelian group G satisfying $x + z \leqslant y + z$ whenever x, y, z belong to G and $x \leqslant y$, and if we set $G^+ = \{x \in G : x \geqslant 0\}$, then (G, G^+) will satisfy (5.1) (i) and (ii).

Definition 5.1.4. For a C^*-algebra A the *positive cone* of $K_0(A)$ is

$$K_0(A)^+ = \{[p]_0 : p \in \mathcal{P}_\infty(A)\} \subseteq K_0(A).$$

Proposition 5.1.5. *Let A be a C^*-algebra.*

(i) $K_0(A)^+ + K_0(A)^+ \subseteq K_0(A)^+$.

(ii) *If A is unital, then $K_0(A)^+ - K_0(A)^+ = K_0(A)$.*

(iii) *If A is stably finite, then $K_0(A)^+ \cap (-K_0(A)^+) = \{0\}$.*

(iv) *If A is unital and stably finite, then $(K_0(A), K_0(A)^+)$ is an ordered Abelian group.*

Proof. (i). If p, q are projections in $\mathcal{P}_\infty(A)$, then $[p]_0 + [q]_0 = [p \oplus q]_0$ belongs to $K_0(A)^+$.

(ii). By the standard picture of $K_0(A)$ for a unital C^*-algebra (Proposition 3.1.7) every element of $K_0(A)$ has the form $[p]_0 - [q]_0$ for some projections p, q in $\mathcal{P}_\infty(A)$.

(iii). Suppose that g is an element in $K_0(A)^+ \cap (-K_0(A)^+)$. Then $g = [p]_0 = -[q]_0$ for some projections p, q in $\mathcal{P}_\infty(A)$. This implies that $[p \oplus q]_0 = 0$ in $K_0(A)$ and hence that $[p \oplus q]_0 = 0$ in $K_0(\widetilde{A})$. By Proposition 3.1.7 there is a projection r in $\mathcal{P}_\infty(\widetilde{A})$ with $p \oplus q \oplus r \sim_0 r$. The projection $p \oplus q \oplus r$ belongs to $\mathcal{P}_n(\widetilde{A})$ for a suitable integer n, and in $\mathcal{P}_n(\widetilde{A})$ there are mutually orthogonal projections p', q', r' with $p \sim_0 p'$, $q \sim_0 q'$, and $r \sim_0 r'$.

Now, $p' + q' + r'$ is Murray–von Neumann equivalent to r' in $M_n(\widetilde{A})$, and $M_n(\widetilde{A})$ is assumed to be finite so that $p' + q' + r'$ is a finite projection. Therefore $p' + q' = 0$. Hence $p' = q' = 0$, $p = q = 0$, and $g = [p]_0 = 0$ in $K_0(A)$.

Part (iv) is an immediate consequence of (i), (ii), and (iii).   □

Let A be a properly infinite, unital C^*-algebra with non-trivial K_0-group; the Cuntz algebras $\mathcal{O}_n$ for $n \geqslant 3$ are examples of such algebras (see Exercise 4.5). Then $K_0(A)^+ = K_0(A)$ by Exercise 4.6 (vi), and hence $K_0(A)^+ \cap -K_0(A)^+ \neq \{0\}$. This shows that $(K_0(A), K_0(A)^+)$ is not an ordered Abelian group.

Definition 5.1.6. An element u in G^+ in an ordered Abelian group (G, G^+) is called an *order unit* if for every g in G there is a positive integer n with $-nu \leqslant g \leqslant nu$.

A triple (G, G^+, u), where (G, G^+) is an ordered Abelian group and u is an order unit, is called an *ordered Abelian group with a distinguished order unit*.

An ordered Abelian group (G, G^+) is said to be *simple* if every non-zero u in G^+ is an order unit.

Not all ordered Abelian groups have order units. For example, let $G = c_0(\mathbf{N}, \mathbf{Z})$ be the additive group of all sequences $x = \{x_n\}_{n=1}^\infty$ of integers x_n such that $x_n = 0$ eventually, and let G^+ be the set of those sequences $x = \{x_n\}_{n=1}^\infty$ for which $x_n \geqslant 0$ for all n. Then (G, G^+) is an ordered Abelian group and no element in G^+ is an order unit.

Proposition 5.1.7. *If A is a unital C^*-algebra, then $[1_A]_0$ is an order unit for $K_0(A)$.*

Proof. Let g in $K_0(A)$ be given. By the standard picture of K_0 of a unital C^*-algebra we can find projections p, q in $\mathcal{P}_n(A)$ for some n such that $g =$

$[p]_0 - [q]_0$. Let 1_n denote the unit of $M_n(A)$, and write 1 instead of 1_A. Then $[1_n]_0 = n[1]_0$, and $1_n - p, 1_n - q$ belong to $\mathcal{P}_n(A)$. Hence

$$-n[1]_0 = -[1_n]_0 = -[q]_0 - [1_n - q]_0 \leqslant -[q]_0 \leqslant [p]_0 - [q]_0$$
$$= g \leqslant [p]_0 \leqslant [p]_0 + [1_n - p]_0 = [1_n]_0 = n[1]_0.$$

$\square$

5.1.8 Positive group homomorphisms. Let (G, G^+) and (H, H^+) be ordered Abelian groups. A group homomorphism $\alpha\colon G \to H$ is said to be *positive* if $\alpha(G^+)$ is contained in H^+, and α is said to be an *order isomorphism* if α is a group isomorphism and $\alpha(G^+) = H^+$.

Assume now that (G, G^+, u) and (H, H^+, v) are ordered Abelian groups with distinguished order units. A positive group homomorphism $\alpha\colon G \to H$ is said to be *order unit preserving* if $\alpha(u) = v$, and the triples (G, G^+, u) and (H, H^+, v) are said to be isomorphic if there exists an order unit preserving order isomorphism from G onto H.

Let A and B be C^*-algebras, and let $\varphi\colon A \to B$ be a *-homomorphism. We know that φ induces a group homomorphism $K_0(\varphi)\colon K_0(A) \to K_0(B)$ (see Paragraph 4.1.2). Since $K_0(\varphi)([p]_0) = [\varphi(p)]_0$ for every projection p in $\mathcal{P}_\infty(A)$, we see that $K_0(\varphi)(K_0(A)^+)$ is contained in $K_0(B)^+$. We refer to this fact by saying that $K_0(\varphi)$ is a *positive group homomorphism* (even in the case where $(K_0(A), K_0(A)^+)$ is not an ordered Abelian group).

If $\varphi\colon A \to B$ is an isomorphism, then $K_0(\varphi)\colon K_0(A) \to K_0(B)$ is a group isomorphism (by functoriality of K_0), and $K_0(\varphi)(K_0(A)^+) = K_0(B)^+$. Hence $K_0(\varphi)$ is an *order isomorphism*. The pair $(K_0(A), K_0(A)^+)$ is therefore an *isomorphism invariant* of A. Moreover, if A is unital, then $K_0(\varphi)([1_A]_0) = [1_B]_0$, and the triple $(K_0(A), K_0(A)^+, [1_A]_0)$ is an isomorphism invariant for A.

An important feature of an isomorphism invariant is that it can tell non-isomorphic C^*-algebras apart. In particular, if A and B are C^*-algebras such that $(K_0(A), K_0(A)^+)$ and $(K_0(B), K_0(B)^+)$ are non-isomorphic, then A is not isomorphic to B. In analogy, the 2-sphere S^2 is not homeomorphic to the 2-torus $\mathbb{T}^2$ because they have different genera (and different fundamental groups, and different K-theories!). Exercise 5.8 at the end of this chapter contains an example of how ordered K-theory can be used to distinguish C^*-algebras.

There is little machinery available to calculate the order structure of $K_0(A)$ for an arbitrary C^*-algebra A. However, we do have the following rather obvious result.

Proposition 5.1.9. *The positive cone of $K_0(A \oplus B)$ is given by*

$$K_0(A \oplus B)^+ = K_0(A)^+ \oplus K_0(B)^+$$
$$(= \{(g,h) : g \in K_0(A)^+, \; h \in K_0(B)^+\}),$$

where $K_0(A \oplus B)$ is identified with $K_0(A) \oplus K_0(B)$ via the isomorphism from Proposition 4.3.4.

Proof. In the notation of Proposition 4.3.4 we must show that

$$(K_0(\iota_A) \oplus K_0(\iota_B))(K_0(A)^+ \oplus K_0(B)^+) = K_0(A \oplus B)^+.$$

Let (g,h) in $K_0(A) \oplus K_0(B)$ be given, and put $x = (K_0(\iota_A) \oplus K_0(\iota_B))(g,h)$. Assume first that x belongs to $K_0(A \oplus B)^+$. Because $K_0(\pi_A)$ and $K_0(\pi_B)$ are positive maps, we get

$$g = K_0(\pi_A)(x) \in K_0(A)^+, \qquad h = K_0(\pi_B)(x) \in K_0(B)^+.$$

Conversely, if g is an element in $K_0(A)^+$ and h is an element in $K_0(B)^+$, then

$$x = K_0(\iota_A)(g) + K_0(\iota_B)(h) \in K_0(A \oplus B)^+$$

because the maps $K_0(\iota_A)$ and $K_0(\iota_B)$ are positive. $\square$

5.2* States on $K_0(A)$ and traces on A

Suppose that (G, G^+, u) is an ordered Abelian group with a distinguished order unit. A *state* on (G, G^+, u) is a group homomorphism $f \colon G \to \mathbb{R}$ satisfying $f(G^+) \subseteq \mathbb{R}^+$ and $f(u) = 1$, i.e., a state is an order unit preserving positive group homomorphism from (G, G^+, u) to $(\mathbb{R}, \mathbb{R}^+, 1)$. The set of all states on (G, G^+, u) is denoted by $S(G, G^+, u)$ or just $S(G)$.

The order structure on G can partly be recovered from its state space as the following Hahn–Banach type theorem by K. Goodearl and D. Handelman shows. (We refer to [22, Theorem 4.12] for its proof.)

Theorem 5.2.1. *Let (G, G^+, u) be an ordered Abelian group with a distinguished order unit, and let x be an element in G. Then $f(x) > 0$ for every f in $S(G)$ if and only if nx is an order unit for (G, G^+) for some positive integer n.*

It is an immediate consequence of this result that if x, y are elements in G with

$f(x) < f(y)$ for all states f on G, then $nx \leqslant ny$ for some natural number n. However, one cannot conclude from this that $x \leqslant y$, see Exercise 5.5.

A positive trace τ on a unital C^*-algebra A induces a positive group homomorphism $K_0(\tau) \colon K_0(A) \to \mathbb{R}$, see Paragraph 3.3.1. To see that $K_0(\tau)$ is positive, note that $K_0(\tau)([p]_0) = \tau(p) \geqslant 0$ for every projection p in $\mathcal{P}_\infty(A)$. If τ is a tracial state, then $K_0(\tau)$ is a state on $(K_0(A), K_0(A)^+, [1_A]_0)$. The result below (proved in [5] and [23]) says — with some technical modifications — that *every* state on $(K_0(A), K_0(A)^+, [1_A]_0)$ comes from a trace in this way.

Theorem 5.2.2. *Every state on* $(K_0(A), K_0(A)^+, [1_A]_0)$, *where A is a unital C^*-algebra, is of the form $K_0(\tau)$ for some* quasi-trace τ *on A.*

If A is exact, *then τ is a trace.*

A quasi-trace on a C^*-algebra A is a continuous function $\tau \colon A^+ \to \mathbb{R}^+$ that satisfies $\tau(x^*x) = \tau(xx^*)$ for all x in A, and $\tau(a + b) = \tau(a) + \tau(b)$ for each pair of *commuting* elements a, b in A^+. The second part of Theorem 5.2.2 follows from a theorem of U. Haagerup [23], saying that every quasi-trace on an exact, unital C^*-algebra extends to a trace.

Most "nice" C^*-algebras are exact. Exactness is more precisely defined in terms of the minimal tensor product as follows: A C^*-algebra B is exact if the functor $A \mapsto A \otimes B$, from the category of C^*-algebras into itself, is exact. We shall not pursue these technical concepts any further in this book.

5.3 Exercises

Exercise 5.1. Let a be an element in a unital C^*-algebra A. Show that a is left-invertible if and only if a^*a is invertible, and that a is right-invertible if and only if aa^* is invertible. [Hint: Look at the proof of Lemma 5.1.2.]

Exercise 5.2. A trace τ on a C^*-algebra A is called *faithful* if $\tau(a) > 0$ for every non-zero positive element a in A.

Let now τ be a positive trace on A, and let $\tau_n \colon M_n(A) \to \mathbb{C}$ be as defined in Paragraph 3.3.1.

(i) Let $x = (x_{ij})_{i,j=1}^n$ in $M_n(A)$ be given. Show that

$$\tau_n(x^*x) = \sum_{i,j=1}^n \tau(x_{ij}^*x_{ij}).$$

(ii) Use (i) to show that τ_n is positive.

(iii) Use (i) to show that τ_n is faithful if τ is faithful.

(iv) Show that if A is a unital C^*-algebra which admits a faithful positive trace, then A is stably finite.

Note. There are partial converses to the statement in (iv) saying: every unital, stably finite C^*-algebra A admits a quasi-trace (which is a trace if A is exact); every unital, stably finite, separable, exact C^*-algebra admits a faithful trace. (See [4] and [23].)

Exercise 5.3. CW-*complexes* form a class of topological spaces which are built from simplexes. Most nice spaces that we encounter are CW-complexes: each topological manifold is a CW-complex, and spaces such as $\mathbb{R}^n$, $\mathbb{T}^n$, S^n, and $\mathbb{D}$ are CW-complexes. Each CW-complex has a dimension (which is a natural number).

The proposition below is proved in Husemoller's book [25, 9.1.2].

Proposition. *Let X be a compact, connected CW-complex that has dimension $\dim X = n$. Let m be the smallest integer such that $m \geqslant (n-1)/2$. Then for every projection p in $\mathcal{P}_\infty(C(X))$ with $\dim p = k > m$ there is a projection q in $\mathcal{P}_\infty(C(X))$ such that $\dim q = m$ and $p \sim q \oplus 1_{k-m}$.*

Use the proposition to show that $(K_0(C(X)), K_0(C(X))^+)$ is a simple ordered Abelian group when X is a compact, connected CW-complex.

Exercise 5.4. Let n be a natural number. Show that the ordered Abelian group with a distinguished order unit $(K_0(M_n(\mathbb{C})), K_0(M_n(\mathbb{C}))^+, [1]_0)$ is isomorphic to $(\mathbb{Z}, \mathbb{Z}^+, n)$.

Exercise 5.5. A simple ordered Abelian group (G, G^+) is said to be *weakly unperforated* if, whenever x in G has the property that nx belongs to $G^+ \setminus \{0\}$ for at least one positive integer n, x belongs to G^+. (The notion of being weakly unperforated in non-simple ordered groups is more subtle.)

Let G be the Abelian group $\mathbb{Z} \oplus \mathbb{Z}$, and put

$$G^+ = \{(k, l) \in G : k \geqslant 2\} \cup \{(0, 0), (1, 0)\}.$$

Show that (G, G^+) is a simple ordered Abelian group which is not weakly unperforated. Show that $u = (1, 0)$ is an order unit for (G, G^+). Show that there is a non-positive element x in G such that $f(x) > 0$ for every state f on (G, G^+, u) .

Exercise 5.6. Let A be a simple, unital C^*-algebra.

(i) Show that $M_n(A)$ is simple for each n.

(ii) Show that $(K_0(A), K_0(A)^+)$ is a simple ordered Abelian group if A is simple, unital, and stably finite. [Hint: Use (i) and Exercise 4.8.]

Exercise 5.7 (Purely infinite, simple C^*-algebras). We begin with a useful definition. A sub-C^*-algebra B of a C^*-algebra A is called a *hereditary sub-C^*-algebra* of A it has the following property: Whenever a, b are positive elements in A such that $a \leqslant b$ and b belongs to B, then a belongs to B.

If B is a hereditary sub-C^*-algebra of A, if b_1, b_2 belong to B, and if a belongs to A, then $b_1 a b_2$ belongs to B. For each positive element a in A, the closure $\overline{aAa}$ of the set $\{axa : x \in A\}$ is a hereditary sub-C^*-algebra of A. Moreover, if A is separable, then every hereditary sub-C^*-algebra of A is of the form $\overline{aAa}$ for some positive element a in A. These facts about hereditary sub-C^*-algebras can be found in [29, Section 3.2], and they can be used without further explanation in this exercise.

(i) Show that the following conditions are equivalent when A is a simple, unital C^*-algebra which is not isomorphic to $\mathbb{C}$.

    (a) For every non-zero, positive element a in A, there is an element x in A such that $1_A = x^* a x$,

    (b) every non-zero, hereditary sub-C^*-algebra of A contains a projection p such that $p \sim 1_A$,

    (c) every non-zero projection in A is properly infinite, and there is a non-zero projection in every non-zero, hereditary sub-C^*-algebra of A.

[Hint: (a) $\Rightarrow$ (b): Let B be a non-zero hereditary sub-C^*-algebra of A and take a non-zero positive element a in B. Use (a) to find x in A with $1_A = x^* a x$. Put $v = a^{1/2} x$, and show that $p = v v^*$ has the desired properties.

(b) $\Rightarrow$ (c): Show first that 1_A is properly infinite as follows. Take a positive element h in A that does not belong to $\mathbb{C} 1_A$. Show that $\mathrm{sp}(h)$ contains more than one point. Use the isomorphism $C(\mathrm{sp}(h)) \cong C^*(1_A, h)$ to find two non-zero, positive elements a, b in A such that $ab = 0$. Take projections p in $\overline{aAa}$ and q in $\overline{bAb}$ with $1_A \sim p \sim q$. Show that $pq = 0$, and conclude that 1_A is properly infinite.

Show next that $1_A \precsim p \precsim 1_A$ for every non-zero projection p in A, and use Exercise 4.7 (iv) to conclude that p is properly infinite.

(c) $\Rightarrow$ (a): Take a non-zero, and hence properly infinite, projection p in $\overline{aAa}$. Find z in A with $\|aza - p\| < 1$. Use Exercise 4.9 (i) to find v in A with $1_A = v^*pv$. Show that v^*aw is invertible when $w = zav$. Use Exercise 5.1 to show that w^*aw is invertible. Finally, find r in A such that $r^*w^*awr = 1_A$.]

A unital, simple C^*-algebra satisfying the equivalent conditions above is called *purely infinite*. Assume in the two questions below that A is a unital, purely infinite, simple C^*-algebra.

(ii) Use Exercise 4.9 to show that if p, q are non-zero projections in A, then $[p]_0 = [q]_0$ if and only if $p \sim q$.

(iii) Combine (ii) with Exercise 4.6 to obtain that the map $[p]_\mathcal{D} \mapsto [p]_0$ restricts to a bijection

$$\{[p]_\mathcal{D} : p \in \mathcal{P}(A), \ p \neq 0\} \to K_0(A),$$

or in other words, the natural map from the set of Murray–von Neumann equivalence classes of non-zero projctions in A into $K_0(A)$ is bijective.

Note. The notion of being purely infinite was introduced by Cuntz in [13]; and he showed in [12] that the Cuntz algebra $\mathcal{O}_n$ is purely infinite and simple for each $n \geqslant 2$. The Calkin algebra $B(H)/\mathcal{K}$, where H is an infinite dimensional, separable Hilbert space, is another example of a purely infinite, simple C^*-algebra. See Exercise 8.13 for more about purely infinite, simple C^*-algebras.

Exercise 5.8 (The irrational rotation C^*-algebras). Let θ be a (fixed) real number, and put $\omega = \exp(2\pi i\theta)$. Consider the Hilbert space $H = L^2(\mathbb{T}^2)$, where $\mathbb{T}^2$ ($= \mathbb{T} \times \mathbb{T}$) is equipped with the normalized Haar measure, and let ξ_0 be the unit vector in H given by $\xi_0(z_1, z_2) = 1$. Define operators u and v on H by

$$(u\,\xi)(z_1, z_2) = z_1\xi(z_1, z_2), \qquad (v\,\xi)(z_1, z_2) = z_2\xi(\omega z_1, z_2), \qquad \xi \in H.$$

(i) Show that u and v are unitary operators, and that $vu = \omega uv$.

Set $A_\theta = C^*(u, v)$, the sub-C^*-algebra of $B(H)$ generated by u and v. Let $\mathcal{A}_\theta$ denote the subset of A_θ that consists of all Laurent polynomials

$$\sum_{n,m \in \mathbb{Z}} \alpha_{n,m} u^n v^m,$$

where only finitely many of the coefficients $\alpha_{n,m}$ are non-zero. Let τ be the positive linear functional on A_θ given by $\tau(a) = \langle a\xi_0, \xi_0 \rangle$.

  (ii) Show that $\mathcal{A}_\theta$ is a sub-*-algebra of $B(H)$, and that $\mathcal{A}_\theta$ is a dense subset of A_θ.

  (iii) Show that $\tau(\sum_{n,m \in \mathbb{Z}} \alpha_{n,m} u^n v^m) = \alpha_{0,0}$ when only finitely many of the coefficients $\alpha_{n,m}$ are non-zero.

  (iv) Show that τ is a tracial state on A_θ. [Hint: Use (iii) to show that $\tau(x^*x) = \tau(xx^*)$ for all x in $\mathcal{A}_\theta$, and then use (ii) and Exercise 3.6.]

The C^*-algebra A_θ is called *the rotation C^*-algebra* corresponding to the angle θ. It has the following *universal property*: if D is any unital C^*-algebra containing unitary elements u', v' satisfying the commutation relation $v'u' = \omega u'v'$, then there is one and only one *-homomorphism $\varphi \colon A_\theta \to D$ with $\varphi(u) = u'$ and $\varphi(v) = v'$.

If θ is irrational, then A_θ is simple and τ is the only tracial state on A_θ. The C^*-algebra A_θ is called an *irrational rotation C^*-algebra* when θ is irrational.

See the notes at the end of the exercise for references to these facts. The reader can use these facts when answering the questions below.

Suppose in the following that θ is an irrational number in the open interval $(0, 1)$. Let $\varphi \colon \mathbb{T} \to \mathbb{T}$ be given by $\varphi(z) = \omega z$. The map φ rotates the circle by angle θ. Let $f, g \colon \mathbb{T} \to \mathbb{R}$ be continuous functions, and set

$$p = f(u)v^* + g(u) + vf(u) \in A_\theta.$$

  (v) Show that $p = p^*$, and that

$$\tau(p) = \int_{\mathbb{T}} g(z)\, dz,$$

where dz is the normalized Haar measure on $\mathbb{T}$. [Hint: Approximate f and g with Laurent polynomials.]

  (vi) Show that $vh(u) = (h \circ \varphi)(u)v$ for every continuous function $h \colon \mathbb{T} \to \mathbb{C}$. [Hint: Approximate h with Laurent polynomials.]

  (vii) Show that $p = p^2$ if and only if

$$f \cdot (f \circ \varphi) = 0, \qquad f \cdot (g + g \circ \varphi^{-1}) = f, \qquad g = g^2 + f^2 + (f \circ \varphi)^2.$$

(viii) Choose ε such that $0 < \varepsilon \leqslant \theta < \theta + \varepsilon \leqslant 1$. Set

$$
g(e^{2\pi it}) = \begin{cases} \varepsilon^{-1}t, & 0 \leqslant t \leqslant \varepsilon, \\ 1, & \varepsilon \leqslant t \leqslant \theta, \\ \varepsilon^{-1}(\theta + \varepsilon - t), & \theta \leqslant t \leqslant \theta + \varepsilon, \\ 0, & \theta + \varepsilon \leqslant t \leqslant 1, \end{cases}
$$

for $t \in [0, 1]$. Show that for this choice of g one can find f such that p is a projection, and show that $\tau(p) = \theta$. [Hint: Use (v) and (vii).]

(ix) Show that

(5.2) $(\mathbb{Z} + \theta\mathbb{Z}) \cap [0, 1] \subseteq \{\tau(e) : e \in \mathcal{P}(A_\theta)\}$.

[Hint: Show that $vu^n = \omega^n u^n v$ for each integer n. For a fixed integer n, let θ' be the (unique) number satisfying $0 < \theta' < 1$ and $\theta' + k = n\theta$ for some integer k. Use (vii) to show that $C^*(u^n, v)$ contains a projection q with $\tau(q) = \theta'$.]

(x) Show that the set $\{\tau(e) : e \in \mathcal{P}(A_\theta)\}$ is countable for each θ, and use this fact and (ix) to show that there is an uncountable set $\mathbb{I}$ of irrational numbers such that the C^*-algebras $\{A_\theta\}_{\theta \in \mathbb{I}}$ are pairwise non-isomorphic.

Notes.   The rotation C^*-algebra A_θ can alternatively be defined to be the *crossed product* $C(\mathbb{T}) \times_\alpha \mathbb{Z}$, where α is the automorphism on $C(\mathbb{T})$ given by $\alpha(f) = f \circ \varphi$. The universal property of A_θ claimed in this exercise then follows from the universal property of crossed products. Simplicity of A_θ and uniqueness of its trace when θ is irrational can be derived from facts about crossed products. See [15], [20], and [30] for details and more information.

The rotation C^*-algebra A_0, corresponding to angle $\theta = 0$, is the universal C^*-algebra generated by two commuting unitaries; and this C^*-algebra is isomorphic to $C(\mathbb{T}^2)$. The non-commutative rotation C^*-algebras A_θ, corresponding to non-integer angles θ, are often called *non-commutative two-tori*.

The projection from question (viii) was constructed by M. Rieffel in [36]; R. Powers showed independently that the irrational rotation C^*-algebras do admit non-trivial projections. M. Pimsner and D. Voiculescu showed subsequently in [33] that the two sets in (5.2) are in fact equal. Moreover, $K_0(\tau)(K_0(A_\theta)) = \mathbb{Z} + \theta\mathbb{Z}$, and the map $K_0(\tau): K_0(A_\theta) \to \mathbb{Z} + \theta\mathbb{Z}$ is an order isomorphism (when $(\mathbb{Z} + \theta\mathbb{Z})^+$ is defined to be $(\mathbb{Z} + \theta\mathbb{Z}) \cap \mathbb{R}^+$). It follows in particular that A_θ is isomorphic to $A_{\theta'}$ if and only if one of the numbers $\theta - \theta'$ and $\theta + \theta'$ is an integer.

Chapter 6

Inductive Limit C^*-Algebras

An inductive limit of a sequence of C^*-algebras is an analogue of a limit of a sequence of points in a topological space. More specifically, an inductive limit of an increasing sequence of sub-C^*-algebras of a given C^*-algebra is the closure of the union of the C^*-algebras in the sequence. Forming inductive limits is a way of making new C^*-algebras from old ones.

We define abstractly inductive limits in arbitrary categories, and it is shown that inductive limits exist in the category of C^*-algebras and in the category of (ordered) Abelian groups. The main result of this chapter says that K_0 is continuous in the sense that forming inductive limits commutes with the functor K_0.

6.1   Products and sums of C^*-algebras

To a family $\{A_i\}_{i \in \mathbb{I}}$ of C^*-algebras we associate two new C^*-algebras, the product $\prod_{i \in \mathbb{I}} A_i$ and the sum $\sum_{i \in \mathbb{I}} A_i$, as follows. Let $\prod_{i \in \mathbb{I}} A_i$ be the set of all functions $a \colon \mathbb{I} \to \bigcup_{i \in \mathbb{I}} A_i$ for which $a(i)$ belongs to A_i for all i in $\mathbb{I}$ and where

$$\|a\| = \sup\{\|a(i)\|_{A_i} : i \in \mathbb{I}\}$$

is finite. We shall hereafter omit the subscript A_i and write $\|a(i)\|$ for the norm of $a(i)$ in A_i. We shall often write $(a_i)_{i \in \mathbb{I}}$, or more briefly (a_i), for the element a in $\prod A_i$ with $a(i) = a_i$.

Equip the set $\prod A_i$ with the operations addition, multiplication, scalar multiplication, and involution by applying these operations coordinate-wise.

Proposition 6.1.1. *The product $\prod_{i \in \mathbb{I}} A_i$ is a C^*-algebra.*

Proof. It is easy to check that $\prod A_i$ is a *-algebra. We show below that the triangle inequality holds for the norm, that $\|a^*a\| = \|a\|^2$, and that $\prod A_i$ is complete.

Let $a = (a_i)_{i \in \mathbb{I}}$ and $b = (b_i)_{i \in \mathbb{I}}$ in $\prod_{i \in \mathbb{I}} A_i$ be given. Observe that $\|a_i\| \leqslant \|a\|$ for all i in $\mathbb{I}$. Now,

$$\|(a+b)_i\| = \|a_i + b_i\| \leqslant \|a_i\| + \|b_i\| \leqslant \|a\| + \|b\|$$

for all i in $\mathbb{I}$, and therefore $\|a + b\| \leqslant \|a\| + \|b\|$.

Since $\|(a^*a)_i\| = \|a_i^* a_i\| = \|a_i\|^2$ for every i in $\mathbb{I}$, we conclude that $\|a^*a\| = \|a\|^2$.

Let $\{a^{(n)}\}_{n=1}^\infty$ be a Cauchy sequence in $\prod A_i$. Then $\{a_i^{(n)}\}_{n=1}^\infty$ is a Cauchy sequence in A_i with a limit a_i in A_i for each fixed i in $\mathbb{I}$. Put $a = (a_i)_{i \in \mathbb{I}}$. Let $\varepsilon > 0$ be given, and find n_0 in $\mathbb{N}$ such that $\|a^{(n)} - a^{(m)}\| \leqslant \varepsilon$ whenever $n, m \geqslant n_0$. Then for each $n \geqslant n_0$ and each i in $\mathbb{I}$,

$$(6.1) \qquad \|a_i^{(n)} - a_i\| = \lim_{m \to \infty} \|a_i^{(n)} - a_i^{(m)}\| \leqslant \varepsilon.$$

Hence

$$\|a_i\| \leqslant \|a_i^{(n_0)}\| + \varepsilon \leqslant \|a^{(n_0)}\| + \varepsilon$$

for all i in $\mathbb{I}$, and this shows that a belongs to $\prod A_i$. It follows from (6.1) that $\|a^{(n)} - a\| \leqslant \varepsilon$ for all $n \geqslant n_0$, proving that $a^{(n)} \to a$. Therefore $\prod A_i$ is complete. $\qquad \square$

Let $\sum_{i \in \mathbb{I}} A_i$ be the closure of the subset

$$\mathcal{I} = \{a \in \prod_{i \in \mathbb{I}} A_i : a(i) = 0 \text{ for all but finitely many } i \in \mathbb{I}\}$$

of $\prod A_i$. The next proposition is easily verified.

Proposition 6.1.2. *The set $\mathcal{I}$ is a (not necessarily closed) two-sided ideal in $\prod_{i \in \mathbb{I}} A_i$, and $\sum_{i \in \mathbb{I}} A_i$ is a closed two-sided ideal in $\prod_{i \in \mathbb{I}} A_i$. In particular, $\sum_{i \in \mathbb{I}} A_i$ is a C*-algebra.*

Let π denote the quotient mapping $\prod A_i \to \prod A_i / \sum A_i$. We have the following useful lemma describing the situation when $\mathbb{I} = \mathbb{N}$.

Lemma 6.1.3. *Let* $\{A_n\}_{n=1}^{\infty}$ *be a sequence of* C^**-algebras, and let* $a = (a_n)_{n \in \mathbb{N}}$ *be an element in* $\prod_{n \in \mathbb{N}} A_n$. *Then*

$$\|\pi(a)\| = \limsup_{n \to \infty} \|a_n\|.$$

In particular, a belongs to $\sum_{n \in \mathbb{N}} A_n$ *if and only if* $\lim_{n \to \infty} \|a_n\| = 0$.

Proof. Since $\mathcal{I}$ is dense in $\sum A_n$, we have $\|\pi(a)\| = \inf\{\|a - b\| : b \in \mathcal{I}\}$ by continuity of the mapping $b \mapsto \|a - b\|$.

Each $b = (b_n)$ in $\mathcal{I}$ has the property that $b_n = 0$ eventually, and therefore

$$\|a - b\| \geqslant \limsup_{n \to \infty} \|a_n - b_n\| = \limsup_{n \to \infty} \|a_n\|.$$

This shows that $\|\pi(a)\| \geqslant \limsup \|a_n\|$.

For each natural number k let $b^{(k)} = (b_n^{(k)})_{n=1}^{\infty}$ be the element in $\mathcal{I}$ given by

$$b_n^{(k)} = \begin{cases} a_n, & n \leqslant k, \\ 0, & n > k. \end{cases}$$

Then

$$\|\pi(a)\| \leqslant \inf_{k \in \mathbb{N}} \|a - b^{(k)}\| = \inf_{k \in \mathbb{N}} \sup_{n > k} \|a_n\| = \limsup_{n \to \infty} \|a_n\|.$$

$\square$

6.2 Inductive limits

It will be convenient to phrase some fundamental definitions and facts about inductive limits in the general context of a category, see Paragraph 3.2.1. Our applications of inductive limits will involve only the categories of C^*-algebras, of Abelian groups, and of ordered Abelian groups (where the morphisms are positive group homomorphisms).

Definition 6.2.1. An *inductive sequence* (or a sequence for short) in a category **C** is a sequence $\{A_n\}_{n=1}^{\infty}$ of objects in **C** and a sequence $\varphi_n \colon A_n \to A_{n+1}$ of morphisms in **C**, usually written

$$A_1 \xrightarrow{\varphi_1} A_2 \xrightarrow{\varphi_2} A_3 \xrightarrow{\varphi_3} \cdots .$$

For $m > n$ we shall also consider the composed morphisms

$$\varphi_{m,n} = \varphi_{m-1} \circ \varphi_{m-2} \circ \cdots \circ \varphi_n : A_n \to A_m,$$

which, together with the morphism φ_n, are called the *connecting morphisms* (or connecting maps). It is sometimes convenient to consider the connecting morphisms $\varphi_{m,n}$ when $m \leqslant n$. They are defined by $\varphi_{n,n} = \mathrm{id}_{A_n}$ and $\varphi_{m,n} = 0$ when $m < n$. (The latter are only defined in categories with a zero object.)

Definition 6.2.2 (Inductive limits). An *inductive limit* of the inductive sequence

(6.2) $$A_1 \xrightarrow{\varphi_1} A_2 \xrightarrow{\varphi_2} A_3 \xrightarrow{\varphi_3} \cdots$$

in a category **C** is a system $(A, \{\mu_n\}_{n=1}^{\infty})$, where A is an object in **C**, where $\mu_n : A_n \to A$ is a morphism in **C** for each n in **N**, and where the following two conditions hold.

(i) The diagram

(6.3)

commutes for each n in **N**.

(ii) If $(B, \{\lambda_n\}_{n=1}^{\infty})$ is a system, where B is an object in **C**, $\lambda_n : A_n \to B$ is a morphism in **C** for each n in **N**, and where $\lambda_n = \lambda_{n+1} \circ \varphi_n$ for all n in **N**, then there is one and only one morphism $\lambda : A \to B$ making the diagram

(6.4)

commutative for each n in **N**.

Inductive limits do not exist in all categories. The (admittedly slightly artificial) category of *finite* sets, where morphisms are functions, does not admit inductive limits, see Exercise 6.4.

Inductive limits, when they exist, are essentially unique in the sense that if $(A, \{\mu_n\})$ and $(B, \{\lambda_n\})$ are both inductive limits of the sequence (6.2), then there is one (and only one) *isomorphism* $\lambda \colon A \to B$ making the diagram (6.4) above commutative.

Indeed, by Definition 6.2.2 (ii) there are unique morphisms $\lambda \colon A \to B$ and $\mu \colon B \to A$ making the left-hand diagram in

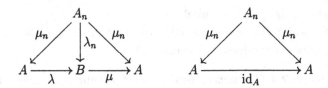

commutative. By the uniqueness clause in Definition 6.2.2 (ii) applied to the two diagrams above we find that $\mu \circ \lambda$ must be the identity map on A. Similarly, $\lambda \circ \mu = \mathrm{id}_B$, and this shows that λ (and μ) are isomorphisms.

Because of the essential uniqueness of inductive limits (when they exist), we shall refer to *the* inductive limit (rather than *an* inductive limit).

The inductive limit of the sequence (6.2) is denoted by $\varinjlim(A_n, \varphi_n)$, or more briefly by $\varinjlim A_n$. We shall also write

$$A_1 \xrightarrow{\varphi_1} A_2 \xrightarrow{\varphi_2} A_3 \xrightarrow{\varphi_3} \cdots \longrightarrow A$$

to indicate that A is the inductive limit of the sequence (6.2).

Example 6.2.3. The following specific examples may illustrate what inductive limits look like.

(i). Let D be a C^*-algebra and let $\{A_n\}_{n=1}^\infty$ be an increasing sequence of sub-C^*-algebras of D. Put

$$A = \overline{\bigcup_{n=1}^\infty A_n},$$

and for each n let $\iota_n \colon A_n \to A$ be the inclusion map. Then $(A, \{\iota_n\})$ is the inductive limit of the sequence $A_1 \to A_2 \to A_3 \to \cdots$ where the connecting maps are the inclusion maps. (See Exercise 6.2.)

(ii). Consider the sequence

$$\mathbb{C} \xrightarrow{\varphi_1} M_2(\mathbb{C}) \xrightarrow{\varphi_2} M_3(\mathbb{C}) \xrightarrow{\varphi_3} \cdots ,$$

where the connecting map φ_n maps an $n \times n$ matrix into the upper left corner of an $(n+1) \times (n+1)$ matrix whose last row and last column are zero. The inductive limit of this sequence is isomorphic to $\mathcal{K}$, the C^*-algebra of compact operators on a separable infinite dimensional Hilbert space. (See Section 6.4.)

(iii). In the category of Abelian groups, the inductive limit of the sequence

$$\mathbb{Z} \xrightarrow{\ 1\ } \mathbb{Z} \xrightarrow{\ 2\ } \mathbb{Z} \xrightarrow{\ 3\ } \cdots \,,$$

is isomorphic to $\mathbb{Q}$. (See Exercise 6.1.)

(iv). A specific example of an inductive limit in the category of ordered Abelian groups is considered in Exercise 6.9.

Proposition 6.2.4 (Inductive limits of C^*-algebras). *Every inductive sequence of C^*-algebras*

$$A_1 \xrightarrow{\ \varphi_1\ } A_2 \xrightarrow{\ \varphi_2\ } A_3 \xrightarrow{\ \varphi_3\ } \cdots$$

has an inductive limit $(A, \{\mu_n\})$. In addition, the following hold.

(i) $A = \overline{\bigcup_{n=1}^{\infty} \mu_n(A_n)}$;

(ii) $\|\mu_n(a)\| = \lim_{m \to \infty} \|\varphi_{m,n}(a)\|$ *for all n in $\mathbb{N}$ and a in A_n;*

(iii) $\mathrm{Ker}(\mu_n) = \{a \in A_n \mid \lim_{m \to \infty} : \varphi_{m,n}(a)\| = 0\}$;

(iv) *if $(B, \{\lambda_n\})$ and $\lambda \colon A \to B$ are as in Definition 6.2.2 (ii), then*

(a) $\mathrm{Ker}(\mu_n) \subseteq \mathrm{Ker}(\lambda_n)$ *for all n in $\mathbb{N}$,*

(b) *λ is injective if and only if $\mathrm{Ker}(\lambda_n) \subseteq \mathrm{Ker}(\mu_n)$ for all n in $\mathbb{N}$,*

(c) *λ is surjective if and only if $B = \overline{\bigcup_{n=1}^{\infty} \lambda_n(A_n)}$.*

Proof. Let

$$\pi \colon \prod_{n=1}^{\infty} A_n \to \prod_{n=1}^{\infty} A_n \Big/ \sum_{n=1}^{\infty} A_n$$

be the quotient mapping (see Section 6.1) and let $\varphi_{m,n} \colon A_n \to A_m$ be the connecting maps associated with the given sequence of C^*-algebras. For each a in A_n, set $\nu_n(a) = (\varphi_{m,n}(a))_{m=1}^{\infty}$ (recalling that $\varphi_{n,n}(a) = a$ and that $\varphi_{m,n}(a) = 0$ when $m < n$), and set $\mu_n = \pi \circ \nu_n$. Then

$$\nu_n \colon A_n \to \prod_{m=1}^{\infty} A_m, \qquad \mu_n \colon A_n \to \prod_{m=1}^{\infty} A_m \Big/ \sum_{m=1}^{\infty} A_m$$

are *-homomorphisms.

Let a be an element in A_n. Then $\nu_n(a) - (\nu_{n+1} \circ \varphi_n)(a)$ is equal to the sequence $(c_m)_{m=1}^{\infty}$, where $c_n = a$ and $c_m = 0$ when $m \neq n$, and this sequence belongs to $\sum_{m=1}^{\infty} A_m$. Hence

$$\mu_n(a) - (\mu_{n+1} \circ \varphi_n)(a) = \pi(\nu_n(a) - (\nu_{n+1} \circ \varphi_n)(a)) = 0,$$

so that $\mu_{n+1} \circ \varphi_n = \mu_n$. It follows from this that $\{\mu_n(A_n)\}_{n=1}^{\infty}$ is an increasing sequence of C^*-algebras. Hence

$$A = \overline{\bigcup_{n=1}^{\infty} \mu_n(A_n)}$$

is a C^*-algebra, (the co-restriction of) each μ_n is a *-homomorphism from A_n to A, and the system $\{\mu_n\}_{n=1}^{\infty}$ satisfies Definition 6.2.2 (i).

Part (i) of the present proposition holds by the construction of A. We now prove (ii). Let a be an element in A_n. Then $\mu_n(a) = \pi(\nu_n(a))$, where $\nu_n(a)$ is the sequence $\{\varphi_{m,n}(a)\}_{m=1}^{\infty}$. Use the norm result in Lemma 6.1.3 to obtain

$$\|\mu_n(a)\| = \|\pi(\nu_n(a))\| = \limsup_{m \to \infty} \|\varphi_{m,n}(a)\| = \lim_{m \to \infty} \|\varphi_{m,n}(a)\|;$$

the limit exists because the sequence $\{\|\varphi_{m,n}(a)\|\}_{m=n}^{\infty}$ is decreasing. It is clear that (iii) follows from (ii).

We proceed to prove (iv)(a) and that Definition 6.2.2 (ii) holds. Assume that $(B, \{\lambda_n\})$ is a system as in Definition 6.2.2 (ii). Then $\lambda_m \circ \varphi_{m,n} = \lambda_n$ for each $m > n$, whence $\|\lambda_n(a)\| \leqslant \|\varphi_{m,n}(a)\|$, and so

$$\|\lambda_n(a)\| \leqslant \limsup_{m \to \infty} \|\varphi_{m,n}(a)\| = \|\mu_n(a)\|.$$

It follows that the kernel of λ_n contains the kernel of μ_n. By the first isomorphism theorem there is a unique *-homomorphism $\lambda_n' \colon \mu_n(A_n) \to B$ such that $\lambda_n' \circ \mu_n = \lambda_n$. By its uniqueness, λ_{n+1}' necessarily extends λ_n'. Hence we obtain a *-homomorphism

$$\lambda' \colon \bigcup_{n=1}^{\infty} \mu_n(A_n) \to B$$

that extends each λ_n'. The map λ' is a *contraction*, i.e., its operator norm is at most 1, because each λ_n' is a contraction. We can therefore extend λ' by uniform continuity to a *-homomorphism $\lambda\colon A \to B$. The restriction of λ to $\mu_n(A_n)$ is equal to λ_n', and hence $\lambda \circ \mu_n = \lambda_n' \circ \mu_n = \lambda_n$. That λ is unique with respect to this property follows from the fact that $\bigcup_{n=1}^{\infty} \mu_n(A_n)$ is dense in A.

(iv)(b). The *-homomorphism λ is injective if and only if it is an isometry, and this is the case if and only if λ' is an isometry, which again is the case if and only if each λ_n' is an isometry, or, equivalently, if and only if each λ_n' is injective. But λ_n' is injective if and only if $\mathrm{Ker}(\lambda_n) = \mathrm{Ker}(\mu_n)$.

(iv)(c). This follow from the fact that the image of λ is $\overline{\bigcup_{n=1}^{\infty} \lambda_n(A_n)}$. □

A similar, but easier, proof shows the following result (see Exercise 6.5).

Proposition 6.2.5 (Inductive limits of Abelian groups). *Each sequence*

$$G_1 \xrightarrow{\ \alpha_1\ } G_2 \xrightarrow{\ \alpha_2\ } G_3 \xrightarrow{\ \alpha_3\ } \cdots$$

of Abelian groups has an inductive limit $(G, \{\beta_n\})$. Moreover,

(i) $G = \bigcup_{n=1}^{\infty} \beta_n(G_n)$,

(ii) $\mathrm{Ker}(\beta_n) = \bigcup_{m=n+1}^{\infty} \mathrm{Ker}(\alpha_{m,n})$ *for each n in $\mathbf{N}$,*

(iii) *if $(H, \{\gamma_n\})$ and $\gamma\colon G \to H$ are as in (ii) of Definition 6.2.2, then*

    (a) *γ is injective if and only if $\mathrm{Ker}(\gamma_n) = \mathrm{Ker}(\beta_n)$ for all n in $\mathbf{N}$,*

    (b) *γ is surjective if and only if $H = \bigcup_{n=1}^{\infty} \gamma_n(G_n)$.*

Proposition 6.2.6 (Inductive limits of ordered Abelian groups). *Let*

(6.5) $$G_1 \xrightarrow{\ \alpha_1\ } G_2 \xrightarrow{\ \alpha_2\ } G_3 \xrightarrow{\ \alpha_3\ } \cdots$$

be a sequence of ordered Abelian groups where the connecting maps are positive group homomorphisms, and let $(G, \{\beta_n\})$ be the inductive limit of this sequence in the category of Abelian groups. Put

$$G^+ = \bigcup_{n=1}^{\infty} \beta_n(G_n^+).$$

Then (G, G^+) is an ordered Abelian group, β_n is a positive group homomorphism for each n in $\mathbf{N}$, and $((G, G^+), \{\beta_n\})$ is the inductive limit of the sequence (6.5) in the category of ordered Abelian groups.

Proof. The connecting maps α_n are positive, i.e., $\alpha_n(G_n^+)$ is contained in G_{n+1}^+ for each n, and therefore

$$\beta_n(G_n^+) = \beta_{n+1}(\alpha_n(G_n^+)) \subseteq \beta_{n+1}(G_{n+1}^+),$$

which shows that $\{\beta_n(G_n^+)\}_{n=1}^{\infty}$ is an increasing sequence of subsets of G. If x, y belong to G^+, then they both belong to $\beta_n(G_n^+)$ for some n, and then $x + y$ is an element in $\beta_n(G_n^+)$ and hence in G^+.

If x is an element in $G^+ \cap -G^+$, then x belongs to $\beta_n(G_n^+) \cap -\beta_n(G_n^+)$ for some n, and hence $x = \beta_n(y_1) = -\beta_n(y_2)$ for some y_1, y_2 in G_n^+. Since $\beta_n(y_1 + y_2) = 0$ there is an integer $m \geqslant n$ such that $\alpha_{m,n}(y_1 + y_2) = 0$. Put $z_j = \alpha_{m,n}(y_j)$. Then z_1, z_2 belong to G_m^+ and $z_1 + z_2 = 0$. This entails that $z_1 = z_2 = 0$ because $G_m^+ \cap -G_m^+ = 0$, and hence $x = \beta_m(z_1) = 0$.

If x is any element in G, then x belongs to $\beta_n(G_n)$ for some n by Proposition 6.2.5 (i). Since $G_n = G_n^+ - G_n^+$, x belongs to $\beta_n(G_n^+) - \beta_n(G_n^+)$ and hence to $G^+ - G^+$. This shows that (G, G^+) is an ordered Abelian group. By the definition of G^+, $\beta_n(G_n^+)$ is contained in G^+ for every n, and so each β_n is positive.

To show that $((G, G^+), \{\beta_n\})$ is the inductive limit of the sequence (6.5) we must verify (i) and (ii) of Definition 6.2.2. That (i) holds follows from the fact that $(G, \{\beta_n\})$ is the inductive limit of the sequence (6.5) *in the category of Abelian groups*. To show (ii), let $(H, \{\gamma_n\})$ be a system, where H is an ordered Abelian group and each $\gamma_n \colon G_n \to H$ is a positive group homomorphism. Using again that G is the inductive limit in the category of Abelian groups, there is a unique group homomorphism $\gamma \colon G \to H$ making the diagram (6.4) commutative. We must show that γ is positive:

$$\gamma(G^+) = \gamma\left(\bigcup_{n=1}^{\infty} \beta_n(G_n^+)\right) = \bigcup_{n=1}^{\infty} (\gamma \circ \beta_n)(G_n^+) = \bigcup_{n=1}^{\infty} \gamma_n(G_n^+) \subseteq H^+.$$

$\square$

6.3 Continuity of K_0

This section contains the main theorem of this chapter (Theorem 6.3.2) which says that the K_0-group of an inductive limit of a sequence of C^*-algebras is the inductive limit of the K_0-groups of the C^*-algebras in the sequence.

Lemma 6.3.1. *Let A be a C*-algebra.*

  (i) *If a is a self-adjoint element in A with $\delta = \|a - a^2\| < 1/4$, then there is a projection p in A with $\|a - p\| \leqslant 2\delta$.*

 (ii) *Let p, q be projections in A. If there exists an element x in A with $\|x^*x - p\| < 1/2$ and $\|xx^* - q\| < 1/2$, then $p \sim q$.*

Proof. (i). If t belongs to $\mathrm{sp}(a)$, then $t - t^2$ belongs to $\mathrm{sp}(a - a^2)$ by the spectral mapping theorem; and if t is a real number satisfying $|t - t^2| \leqslant \delta < 1/4$, then t belongs to $[-2\delta, 2\delta] \cup [1 - 2\delta, 1 + 2\delta]$. Hence, if $\|a - a^2\| = \delta < 1/4$, then

$$\mathrm{sp}(a) \subseteq \{t \in \mathbb{R} : |t - t^2| \leqslant \delta\} \subseteq [-2\delta, 2\delta] \cup [1 - 2\delta, 1 + 2\delta].$$

It follows that we can define a continuous function f on $\mathrm{sp}(a)$ by

$$(6.6) \qquad\qquad f(t) = \begin{cases} 0, & t \leqslant 2\delta, \\ 1, & t \geqslant 1 - 2\delta. \end{cases}$$

Put $p = f(a)$. Then p is a projection because $f = f^2 = \overline{f}$, and $\|a - p\| \leqslant 2\delta$ because $|t - f(t)| \leqslant 2\delta$ for all t in $\mathrm{sp}(a)$.

(ii). Put

$$\delta = \frac{1}{2} \max\{\|x^*x - p\|, \|xx^* - q\|\} < 1/4,$$

and put $\Gamma = \mathrm{sp}(x^*x) \cup \mathrm{sp}(xx^*)$. Then $\Gamma \subseteq [-2\delta, 2\delta] \cup [1 - 2\delta, 1 + 2\delta]$ by Lemma 2.2.3. Let f in $C(\Gamma)$ be as defined in (6.6). Then, as in (i), $p_0 = f(x^*x)$ and $q_0 = f(xx^*)$ are projections in A satisfying $\|p - p_0\| \leqslant 4\delta < 1$ and $\|q - q_0\| \leqslant 4\delta < 1$. Hence $p \sim p_0$ and $q \sim q_0$ by Proposition 2.2.4.

Notice that $xh(x^*x)x^* = h(xx^*)xx^*$ for every h in $C(\Gamma)$; this is clear when h is a polynomial, and the polynomials are dense in $C(\Gamma)$. Let g be the positive function in $C(\Gamma)$ that satisfies $tg(t)^2 = f(t)$ for all t in Γ. Put $v = xg(x^*x)$. Then

$$v^*v = g(x^*x)x^*xg(x^*x) = f(x^*x) = p_0,$$
$$vv^* = xg(x^*x)^2x^* = g(xx^*)^2xx^* = f(xx^*) = q_0.$$

Hence $p_0 \sim q_0$, and it follows that $p \sim q$ as desired. □

Theorem 6.3.2 (Continuity of K_0). *For each inductive sequence*

$$(6.7) \qquad\qquad A_1 \xrightarrow{\varphi_1} A_2 \xrightarrow{\varphi_2} A_3 \xrightarrow{\varphi_3} \cdots$$

of C^*-algebras, $K_0(\varinjlim A_n)$ and $\varinjlim K_0(A_n)$ are isomorphic as Abelian groups. If $(K_0(A_n), K_0(A_n)^+)$ is an ordered Abelian group for each n, then $K_0(\varinjlim A_n)$ is isomorphic to $\varinjlim K_0(A_n)$ in the category of ordered Abelian groups.

More specifically, if $(A, \{\mu_n\})$ is the inductive limit of the sequence (6.7), and if $(G_0, \{\beta_n\})$ is the inductive limit of the sequence

$$(6.8) \qquad K_0(A_1) \xrightarrow{\;K_0(\varphi_1)\;} K_0(A_2) \xrightarrow{\;K_0(\varphi_2)\;} K_0(A_3) \xrightarrow{\;K_0(\varphi_3)\;} \cdots,$$

then there is a unique group isomorphism $\gamma\colon G_0 \to K_0(A)$ that makes the diagram

$$(6.9)$$

$$
\begin{array}{ccc}
 & K_0(A_n) & \\
 {\scriptstyle \beta_n}\nearrow & & \searrow{\scriptstyle K_0(\mu_n)} \\
 G_0 & \xrightarrow{\quad\gamma\quad} & K_0(A)
\end{array}
$$

commutative for each n in $\mathbb{N}$. Moreover,

(i) $K_0(A) = \bigcup_{n=1}^{\infty} K_0(\mu_n)(K_0(A_n))$,

(ii) $K_0(A)^+ = \bigcup_{n=1}^{\infty} K_0(\mu_n)(K_0(A_n)^+)$,

(iii) $\mathrm{Ker}(K_0(\mu_n)) = \bigcup_{m=n+1}^{\infty} \mathrm{Ker}(K_0(\varphi_{m,n}))$ *for each n in $\mathbb{N}$.*

Proof. We first prove (i), (ii), and (iii). To simplify the notation, let μ_n denote the induced $*$-homomorphism $M_k(A_n) \to M_k(A)$ for each k in $\mathbb{N}$, and let $\tilde{\mu}_n$ denote the induced $*$-homomorphism $M_k(\tilde{A}_n) \to M_k(\tilde{A})$ where units have been adjoined to A_n and to A. Using Proposition 6.2.4 (iv) one finds that $(M_k(A), \{\mu_n\})$ and $(M_k(\tilde{A}), \{\tilde{\mu}_n\})$ are the inductive limits of the sequences

$$M_k(A_1) \xrightarrow{\varphi_1} M_k(A_2) \xrightarrow{\varphi_2} M_k(A_3) \xrightarrow{\varphi_3} \cdots,$$

$$M_k(\tilde{A}_1) \xrightarrow{\tilde{\varphi}_1} M_k(\tilde{A}_2) \xrightarrow{\tilde{\varphi}_2} M_k(\tilde{A}_3) \xrightarrow{\tilde{\varphi}_3} \cdots.$$

(i). Let g in $K_0(A)$ be given. Use the standard picture of $K_0(A)$ to find a natural number k and a projection p in $M_k(\tilde{A})$ with $g = [p]_0 - [s(p)]_0$. By Proposition 6.2.4 (i) there exist a natural number n and an element b_n in $M_k(\tilde{A}_n)$ such that $\|\tilde{\mu}_n(b_n) - p\| < 1/5$. Put $a_n = (b_n + b_n^*)/2$, and put $a_m = \tilde{\varphi}_{m,n}(a_n)$ for $m > n$. Then a_m is self-adjoint and $\|\tilde{\mu}_m(a_m) - p\| < 1/5$

for $m \geqslant n$. Lemma 2.2.3 implies that

$$\operatorname{sp}(\widetilde{\mu}_n(a_n)) \subseteq [-1/5, 1/5] \cup [4/5, 6/5],$$

and so

$$\|\widetilde{\mu}_n(a_n^2 - a_n)\| = \max\{|t^2 - t| : t \in \operatorname{sp}(\widetilde{\mu}_n(a_n))\} < 1/4.$$

By Proposition 6.2.4 (ii) there is $m \geqslant n$ such that $\|a_m^2 - a_m\| < 1/4$. By Lemma 6.3.1 (i) there is a projection q in $M_k(\widetilde{A}_m)$ with $\|a_m - q\| < 1/2$. Now,

$$\|\widetilde{\mu}_m(q) - p\| \leqslant \|q - a_m\| + \|\widetilde{\mu}_m(a_m) - p\| < 1,$$

and so $\widetilde{\mu}_m(q) \sim p$. Hence, by the standard picture for K_0 (Proposition 4.2.2 (i) and (iii)),

$$g = [p]_0 - [s(p)]_0 = [\widetilde{\mu}_m(q)]_0 - [s(\widetilde{\mu}_m(q))]_0 = K_0(\mu_m)([q]_0 - [s(q)]_0).$$

(ii). For each n in $\mathbb{N}$, $K_0(\mu_n)(K_0(A_n)^+)$ is contained in $K_0(A)^+$ because $K_0(\mu_n)$ is positive. Conversely, let g in $K_0(A)^+$ be given. Then $g = [p]_0$ for some natural number k and some projection p in $M_k(A)$. Arguing as in (i) we find a natural number m and a projection q in $M_k(A_m)$ such that $\|\mu_m(q) - p\| < 1$. It follows that $p \sim \mu_m(q)$, and hence that $g = [\mu_m(q)]_0 = K_0(\mu_m)([q]_0)$; this proves (ii) because $[q]_0$ belongs to $K_0(A_m)^+$.

(iii). The kernel of $K_0(\mu_n)$ contains the kernel of $K_0(\varphi_{m,n})$ for each $m > n$ by Proposition 6.2.4. Conversely, let g in $K_0(A_n)$ be an element in the kernel of $K_0(\mu_n)$. By the appendix to the standard picture for K_0 (Lemma 4.2.3 (i)) there are k in $\mathbb{N}$ and a projection p in $M_k(\widetilde{A}_n)$ such that $g = [p]_0 - [s(p)]_0$ and $\widetilde{\mu}_n(p) \sim \widetilde{\mu}_n(s(p))$. In other words, $\widetilde{\mu}_n(p) = v^*v$ and $\widetilde{\mu}_n(s(p)) = vv^*$ for some v in $M_k(\widetilde{A})$. By Proposition 6.2.4 (i) there exist an integer $l \geqslant n$ and an element x_l in $M_k(\widetilde{A}_l)$ with $\widetilde{\mu}_l(x_l)$ close enough to v to ensure that

$$\|\widetilde{\mu}_l(x_l^* x_l) - \widetilde{\mu}_n(p)\| < 1/2, \qquad \|\widetilde{\mu}_l(x_l x_l^*) - \widetilde{\mu}_n(s(p))\| < 1/2.$$

By Proposition 6.2.4 (iii) there is $m \geqslant l$ such that

$$\|x_m^* x_m - \widetilde{\varphi}_{m,n}(p)\| < 1/2, \qquad \|x_m x_m^* - \widetilde{\varphi}_{m,n}(s(p))\| < 1/2,$$

where $x_m = \widetilde{\varphi}_{m,l}(x_l)$. Lemma 6.3.1 (ii) now yields that

$$\widetilde{\varphi}_{m,n}(p) \sim \widetilde{\varphi}_{m,n}(s(p)) = s(\widetilde{\varphi}_{m,n}(p)) \quad \text{in} \quad M_k(\widetilde{A}_m).$$

This shows that

$$K_0(\varphi_{m,n})(g) = [\widetilde{\varphi}_{m,n}(p)]_0 - [s(\widetilde{\varphi}_{m,n}(p))]_0 = 0,$$

and (iii) follows.

Since $K_0(\mu_n) = K_0(\mu_{n+1}) \circ K_0(\varphi_n)$ for every n, part (ii) of the definition of inductive limits (Definition 6.2.2) applied to the system $(K_0(A), \{K_0(\mu_n)\})$ yields a unique group homomorphism $\gamma \colon G_0 \to K_0(A)$ making the diagram (6.9) commutative for every n. It follows from (i) that γ is surjective.

Suppose that g belongs to the kernel of γ. Find n in $\mathbb{N}$ and h in $K_0(A_n)$ with $g = \beta_n(h)$, see Proposition 6.2.5 (i). Then $0 = \gamma(g) = K_0(\mu_n)(h)$, and by (iii) this implies that $K_0(\varphi_{m,n})(h) = 0$ for some integer $m > n$, whence $g = (\beta_m \circ K_0(\varphi_{m,n}))(h) = 0$. This shows that γ is injective.

Finally, if $(K_0(A_n), K_0(A_n)^+)$ is an ordered Abelian group for each n, then the inductive limit of the sequence (6.8) in the category of ordered Abelian groups is $((G_0, G_0^+), \{\beta_n\})$, where

$$G_0^+ = \bigcup_{n=1}^{\infty} \beta_n(K_0(A_n)^+),$$

see Proposition 6.2.6. Now (ii) shows that $\gamma(G_0^+) = K_0(A)^+$, and so γ is an isomorphism of ordered Abelian groups.   □

6.4* Stabilized C^*-algebras

The stabilization of a C^*-algebra A is the C^*-algebra $\mathcal{K} \otimes A$, where $\mathcal{K}$ is the C^*-algebra of all compact operators on a separable infinite dimensional Hilbert space. The stability property of K-theory should be phrased by saying that the natural *-homomorphism $A \to \mathcal{K} \otimes A$ induces an isomorphism $K_0(A) \to K_0(\mathcal{K} \otimes A)$ (and similarly for K_1).

In order to avoid discussing tensor products, and for the purpose of giving a more useful picture of stabilization, we shall for each C^*-algebra A construct a new C^*-algebra $\mathcal{K}A$ which we shall call *the stabilization of A*. It will be proved that $\mathcal{K}\mathbb{C}$ is isomorphic to $\mathcal{K}$, and the reader who is familiar with tensor products can easily prove that $\mathcal{K}A$ is isomorphic to $\mathcal{K} \otimes A$ for every C^*-algebra A.

Consider the sequence of C^*-algebras

$$A \xrightarrow{\varphi_1} M_2(A) \xrightarrow{\varphi_2} M_3(A) \xrightarrow{\varphi_3} \cdots,$$

where the connecting maps $\varphi_n \colon M_n(A) \to M_{n+1}(A)$ are given by

$$\varphi_n(a) = \begin{pmatrix} a & 0 \\ 0 & 0 \end{pmatrix}, \qquad a \in M_n(A).$$

Let $(\mathcal{K}A, \{\kappa_n\})$ be the inductive limit of this sequence, where $\kappa_n \colon M_n(A) \to \mathcal{K}A$, and set $\kappa = \kappa_1$, so that $\kappa \colon A \to \mathcal{K}A$. The C^*-algebra $\mathcal{K}A$ is our stabilization of A.

We proceed to prove that $\mathcal{K}\mathbb{C} \cong \mathcal{K}$. Choose an orthonormal basis $\{e_n\}_{n=1}^{\infty}$ for H, and let H_n be the subspace of H spanned by $\{e_1, \ldots, e_n\}$. Then $\{H_n\}_{n=1}^{\infty}$ is an increasing sequence of subspaces of H and $\bigcup_{n=1}^{\infty} H_n$ is dense in H. Let F_n in $B(H)$ be the projection onto H_n. Choose an isomorphism $\alpha_n \colon M_n(\mathbb{C}) \to B(H_n)$ such that a is the matrix for the linear map $\alpha_n(a)$ with respect to the basis $\{e_1, \ldots, e_n\}$ for each a in $M_n(\mathbb{C})$. We may identify $B(H_n)$ with the subalgebra $F_n B(H) F_n$ of $B(H)$. It is a standard fact that $\mathcal{K}$ is the closure of the union $\bigcup_{n=1}^{\infty} F_n B(H) F_n$, see [31, Lemma 3.3.2 and Theorem 3.3.3]. With these identifications we get the commutative diagram

$$
\begin{array}{ccccccccc}
\mathbb{C} & \xrightarrow{\varphi_1} & M_2(\mathbb{C}) & \xrightarrow{\varphi_2} & M_3(\mathbb{C}) & \xrightarrow{\varphi_3} & \cdots & \longrightarrow & \mathcal{K}\mathbb{C} \\
\downarrow{\alpha_1} & & \downarrow{\alpha_2} & & \downarrow{\alpha_3} & & & & \vdots\,\alpha \\
F_1 B(H) F_1 & \xrightarrow{\iota} & F_2 B(H) F_2 & \xrightarrow{\iota} & F_3 B(H) F_3 & \xrightarrow{\iota} & \cdots & \longrightarrow & \mathcal{K}.
\end{array}
$$

It follows from Exercise 6.2 that $\mathcal{K}$ is the inductive limit of the sequence in the bottom row of this diagram, where the connecting maps are the inclusion maps. The existence of the isomorphism α indicated with the dashed arrow follows from Exercise 6.8.

Proposition 6.4.1 (Stability of K_0). *Let $\kappa \colon A \to \mathcal{K}A$ be the canonical inclusion of a C^*-algebra A into its stabilization $\mathcal{K}A$ defined above. Then $K_0(\kappa) \colon K_0(A) \to K_0(\mathcal{K}A)$ is an isomorphism.*

Proof. The connecting homomorphism $\varphi_{n,1} \colon A \to M_n(A)$ is equal to the map $\lambda_{n,A}$ considered in Proposition 4.3.8, and so $K_0(\varphi_{n,1}) \colon K_0(A_n) \to K_0(M_n(A))$ is an isomorphism for each n.

Let g be an element in $K_0(\mathcal{K}A)$. Then $g = K_0(\kappa_n)(g')$ for some natural number n and some g' in $K_0(M_n(A))$ by Theorem 6.3.2 (i). Now $g' = K_0(\varphi_{n,1})(h)$ for some h in $K_0(A)$, and hence

$$K_0(\kappa)(h) = K_0(\kappa_n \circ \varphi_{n,1})(h) = K_0(g') = g,$$

showing that $K_0(\kappa)$ is surjective.

Suppose that h is an element in $K_0(A)$ and that $K_0(\kappa)(h) = 0$. Then $K_0(\varphi_{n,1})(h) = 0$ for some $n \geq 2$ by Theorem 6.3.2 (iii), and this entails that $h = 0$. Hence $K_0(\kappa)$ is injective. $\square$

Let H be a Hilbert space, and let Tr denote the (unbounded) trace on $B(H)$ given by the formula

$$(6.10) \qquad\qquad \mathrm{Tr}(T) = \sum_{n=1}^{\infty} \langle Te_n, e_n \rangle,$$

where $\{e_n\}_{n=1}^{\infty}$ is (any) orthonormal basis for the Hilbert space H. (See [31, Section 3.3] for more about the trace Tr.) The sum (6.10) is defined for all positive operators and for all trace class operators T. (We allow $\mathrm{Tr}(T) = \infty$ when T is positive.) It is an easy consequence of the fact that Tr is independent of the choice of orthonormal basis $\{e_n\}_{n=1}^{\infty}$ that $\mathrm{Tr}(E) = \dim(E(H))$ for each projection E on H.

Corollary 6.4.2. *The group $K_0(\mathcal{K})$ is isomorphic to $\mathbb{Z}$. More specifically, there is an isomorphism $\alpha\colon K_0(\mathcal{K}) \to \mathbb{Z}$ such that $\alpha([E]_0) = \mathrm{Tr}(E)$ for each projection E in $\mathcal{K}$.*

The isomorphism α above is commonly denoted by $K_0(\mathrm{Tr})$.

Proof of corollary. Identify $\mathcal{K}$ with $\mathcal{K}\mathbb{C}$, and consider the map $\kappa\colon \mathbb{C} \to \mathcal{K}\mathbb{C}$ from the stabilization construction. From Example 3.3.2 we have an isomorphism $\alpha_1\colon K_0(\mathbb{C}) \to \mathbb{Z}$ with $\alpha_1([1]_0) = 1$. Put

$$\alpha = \alpha_1 \circ K_0(\kappa)^{-1}\colon K_0(\mathcal{K}) \to \mathbb{Z}.$$

By the specific identification of $\mathcal{K}$ with $\mathcal{K}\mathbb{C}$ given above, $F = \kappa(1)$ is a one-dimensional projection in $\mathcal{K}$, and consequently

$$\alpha([F]_0) = \alpha_1([1]_0) = 1.$$

If E is an arbitrary one-dimensional projection on H, then $F \sim E$, and hence $\alpha([E]_0) = \alpha([F]_0) = 1$. Finally, if E is an arbitrary n-dimensional projection on H, then E is the sum of n one-dimensional projections, and it follows by additivity of α that $\alpha([E]_0) = n = \text{Tr}(E)$. $\qquad\square$

The stabilized C^*-algebra $\mathcal{K}A$ contains all matrix algebras over A via the embeddings $\kappa_n \colon M_n(A) \to \mathcal{K}A$. This has the interesting consequence that one can define $K_0(A)$ directly using the C^*-algebra $\mathcal{K}A$ instead of the system of all matrix algebras over A. Indeed, if A is a C^*-algebra, then the set of (Murray–von Neumann) equivalence classes of projections in $\mathcal{K}A$ is isomorphic, as an Abelian semigroup, to $\mathcal{D}(A)$, the semigroup of equivalence classes of projections in $\mathcal{P}_\infty(A)$. (See Exercise 6.6.)

6.5 Exercises

Exercise 6.1. Let G_1 and G_2 be the inductive limits of the following two sequences of Abelian groups:

$$\mathbb{Z} \xrightarrow{\ 1\ } \mathbb{Z} \xrightarrow{\ 2\ } \mathbb{Z} \xrightarrow{\ 3\ } \mathbb{Z} \longrightarrow \cdots, \qquad \mathbb{Z} \xrightarrow{\ 2\ } \mathbb{Z} \xrightarrow{\ 2\ } \mathbb{Z} \xrightarrow{\ 2\ } \mathbb{Z} \longrightarrow \cdots,$$

where the homomorphism $n \colon \mathbb{Z} \to \mathbb{Z}$ is given by $1 \mapsto n$. Show that $G_1 \cong \mathbb{Q}$. Determine G_2.

Exercise 6.2. Let D be a C^*-algebra and let $\{A_n\}_{n=1}^\infty$ be an increasing sequence of sub-C^*-algebras of D. Put

$$A = \overline{\bigcup_{n=1}^\infty A_n},$$

and for each n let $\iota_n \colon A_n \to A$ be the inclusion map. Show that $(A, \{\iota_n\})$ is (isomorphic to) the inductive limit of the sequence $A_1 \to A_2 \to A_3 \to \cdots$, where the connecting maps are the inclusion maps.

Exercise 6.3. Let $(A, \{\mu_n\})$ be the inductive limit C^*-algebra of the sequence

$$A_1 \xrightarrow{\ \varphi_1\ } A_2 \xrightarrow{\ \varphi_2\ } A_3 \xrightarrow{\ \varphi_3\ } \cdots.$$

Let a be an element in A_n and let b be an element in A_m. Show that $\mu_n(a) =$

$\mu_m(b)$ if and only if

$$\lim_{k\to\infty} \|\varphi_{k,n}(a) - \varphi_{k,m}(b)\| = 0.$$

Exercise 6.4. Let $\mathbf{Set}_f$ be the category of all finite sets, and for each pair of finite sets X and Y, let $\mathrm{Mor}(X,Y)$ be the set of all functions from X to Y. Let, similarly, $\mathbf{Set}$ be the category of all sets (finite or not).

(i) Let

(6.11) $$X_1 \xrightarrow{f_1} X_2 \xrightarrow{f_2} X_3 \xrightarrow{f_3} \cdots$$

be a sequence in $\mathbf{Set}_f$, where all the functions f_n are injective and where X_n has n elements. Show that for each finite set Y there are functions $g_n\colon X_n \to Y$ such that $g_{n+1} \circ f_n = g_n$ and $Y = \bigcup_{n=1}^{\infty} g_n(X_n)$. Show next that the sequence in (6.11) does not admit an inductive limit in $\mathbf{Set}_f$. Conclude that the category $\mathbf{Set}_f$ does not admit inductive limits.

(ii) Show that the category $\mathbf{Set}$ does admit inductive limits.

Exercise 6.5. Prove Proposition 6.2.5. [Hint: Given a sequence of Abelian groups

$$G_1 \xrightarrow{\alpha_1} G_2 \xrightarrow{\alpha_2} G_3 \xrightarrow{\alpha_3} \cdots,$$

consider the group $\prod_{n=1}^{\infty} G_n$ of all infinite sequences $(g_1, g_2, g_3, \ldots)$, where g_n belongs to G_n, equipped with entry-wise defined group operation, and let $\sum_{n=1}^{\infty} G_n$ be the subgroup consisting of those sequences $(g_1, g_2, g_3, \ldots)$, where $g_n = 0$ eventually. Following the proof of Proposition 6.2.4, construct for each natural number n a suitable group homomorphism β_n from G_n into $\prod_{n=1}^{\infty} G_n / \sum_{n=1}^{\infty} G_n$, put $G = \bigcup_{n=1}^{\infty} \beta_n(G_n)$, and show that $(G, \{\beta_n\})$ is an inductive limit of the given sequence of Abelian groups.]

Exercise 6.6. Let A be a C^*-algebra. Define a map $\varrho\colon \mathcal{P}_\infty(A) \to \mathcal{P}(\mathcal{K}A)$ as follows. If p is an element in $\mathcal{P}_n(A)$, then set $\varrho(p) = \kappa_n(p)$, where $\kappa_n\colon M_n(A) \to \mathcal{K}A$ is as in Section 6.4.

(i) Let p, q be projections in $\mathcal{P}_\infty(A)$ with $\varrho(p) \sim \varrho(q)$. Show that $p \sim_0 q$. [Hint: Look at the proof of Theorem 6.3.2 (iii).]

(ii) Let p be a projection in $\mathcal{P}(\mathcal{K}A)$. Show that there exists a projection q in $\mathcal{P}_\infty(A)$ such that $\varrho(q) \sim p$. [Hint: Look at the proof of Theorem 6.3.2 (i).]

(iii) Conclude from (i) and (ii) that ϱ induces a bijection

$$\hat{\varrho}\colon \mathcal{D}(A) = \mathcal{P}_\infty(A)/\sim_0 \ \to \mathcal{P}(\mathcal{K}A)/\sim.$$

(iv) Let p, q be projections in $\mathcal{P}(\mathcal{K}A)$. Show that there are projections p', q' in $\mathcal{P}(\mathcal{K}A)$ such that $p' \sim p$, $q' \sim q$, and $p' \perp q'$. [Hint: First use (ii) to find a natural number n and projections p_1, q_1 in $\mathcal{P}_n(A)$ such that $\varrho(p_1) \sim p$ and $\varrho(q_1) \sim q$.]

Note. One can turn $\mathcal{P}(\mathcal{K}A)/\sim$ into an Abelian semigroup as follows. Given a pair of projections p, q in $\mathcal{K}A$, use (iv) to find mutually orthogonal projections p', q' in $\mathcal{K}A$ with $p' \sim p$ and $q' \sim q$. Then define the sum of the equivalence classes containing p, and q, to be the equivalence class containing the projection $p' + q'$. The map ϱ above is then an isomorphism of semigroups.

Exercise 6.7. Let

$$A_1 \xrightarrow{\ \varphi_1\ } A_2 \xrightarrow{\ \varphi_2\ } A_3 \xrightarrow{\ \varphi_3\ } \cdots$$

be a sequence of C^*-algebras with inductive limit $(A, \{\mu_n\})$.

(i) Suppose that $1 \leqslant n_1 < n_2 < n_3 < \cdots$, and put $\psi_j = \varphi_{n_{j+1}, n_j}$. Show that $(A, \{\mu_{n_j}\})$ is the inductive limit of the sequence

$$A_{n_1} \xrightarrow{\ \psi_1\ } A_{n_2} \xrightarrow{\ \psi_2\ } A_{n_3} \longrightarrow \cdots .$$

(ii) Put $B_n = A_n/\mathrm{Ker}(\mu_n)$, and let $\pi_n\colon A_n \to B_n$ be the quotient mapping. Justify that there are injective *-homomorphisms $\psi_n\colon B_n \to B_{n+1}$ and a *-homomorphism $\pi\colon A \to \varinjlim B_n$ making the diagram

$$
\begin{array}{ccccccccc}
A_1 & \xrightarrow{\ \varphi_1\ } & A_2 & \xrightarrow{\ \varphi_2\ } & A_3 & \to \cdots & \longrightarrow & A \\
\downarrow{\scriptstyle \pi_1} & & \downarrow{\scriptstyle \pi_2} & & \downarrow{\scriptstyle \pi_3} & & & \downarrow{\scriptstyle \pi} \\
B_1 & \xrightarrow[\ \psi_1\ ]{} & B_2 & \xrightarrow[\ \psi_2\ ]{} & B_3 & \to \cdots & \to & \varinjlim B_n
\end{array}
$$

commutative. Show that π is an isomorphism.

(iii) Suppose that each map $\varphi_n\colon A_n \to A_{n+1}$ is injective. Show that each map $\mu_n\colon A_n \to A$ is injective. Suppose further that A is unital. Show that there exists a natural number n_0 such that for all integers $n \geqslant n_0$, A_n is unital, and the maps $\varphi_n\colon A_n \to A_{n+1}$ and $\mu_n\colon A_n \to A$ are unit preserving.

Exercise 6.8. Let

$$A_1 \xrightarrow{\varphi_1} A_2 \xrightarrow{\varphi_2} A_3 \xrightarrow{\varphi_3} \cdots, \qquad B_1 \xrightarrow{\psi_1} B_2 \xrightarrow{\psi_2} B_3 \xrightarrow{\psi_3} \cdots$$

be two inductive sequences of C^*-algebras. Suppose that there are *-homomorphisms $\alpha_n\colon A_n \to B_n$ and $\beta_n\colon B_n \to A_{n+1}$ making the diagram

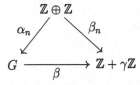

commutative. Show that there are *-isomorphisms α and β as indicated with the dashed arrows in the diagram making the entire diagram commutative. In particular, A is isomorphic to B.

Exercise 6.9. Consider the inductive sequence of ordered Abelian groups

$$\mathbb{Z} \oplus \mathbb{Z} \xrightarrow{\varphi} \mathbb{Z} \oplus \mathbb{Z} \xrightarrow{\varphi} \mathbb{Z} \oplus \mathbb{Z} \xrightarrow{\varphi} \cdots,$$

where $\mathbb{Z} \oplus \mathbb{Z}$ is equipped with the standard ordering given by

$$(\mathbb{Z} \oplus \mathbb{Z})^+ = \{(x,y) : x \geqslant 0,\ y \geqslant 0\},$$

and where $\varphi(x,y) = (x+y,x)$.

Let $(G, \{\alpha_n\})$ be the inductive limit of this sequence. Let γ be the golden ratio $(1+\sqrt{5})/2$. The purpose of this exercise is to show that G is isomorphic to the ordered Abelian group $\mathbb{Z} + \gamma\mathbb{Z}$ (whose positive cone is $(\mathbb{Z}+\gamma\mathbb{Z})\cap\mathbb{R}^+$). Let $\beta_n\colon \mathbb{Z} \oplus \mathbb{Z} \to \mathbb{Z} + \gamma\mathbb{Z}$ be given by $\beta_n(x,y) = \gamma^{2-n}x + \gamma^{1-n}y$.

(i) Show that there exists a positive group homomorphism $\beta\colon G \to \mathbb{Z}+\gamma\mathbb{Z}$ such that the diagram

$$
\begin{array}{ccc}
 & \mathbb{Z} \oplus \mathbb{Z} & \\
{\scriptstyle\alpha_n}\swarrow & & \searrow{\scriptstyle\beta_n} \\
G & \xrightarrow{\quad\beta\quad} & \mathbb{Z}+\gamma\mathbb{Z}
\end{array}
$$

is commutative for each n in $\mathbb{N}$, and show that β is bijective.

(ii) Find non-zero elements u and v in $\mathbb{R} \oplus \mathbb{R}$ such that

$$\begin{pmatrix} 1 & 1 \\ 1 & 0 \end{pmatrix} u = \gamma u, \qquad \begin{pmatrix} 1 & 1 \\ 1 & 0 \end{pmatrix} v = -\gamma^{-1} v,$$

$\beta_1(u) > 0$, and $\beta_1(v) = 0$.

(iii) Let t be in $\mathbb{Z} + \gamma\mathbb{Z}$, and assume that $t > 0$. For each n let g_n be the element in $\mathbb{Z} \oplus \mathbb{Z}$ with $\beta_n(g_n) = t$. Let a, b be real numbers satisfying $g_1 = au + bv$. Show that $a > 0$ and that

$$g_n = \begin{pmatrix} 1 & 1 \\ 1 & 0 \end{pmatrix}^{n-1} g_1.$$

Show that g_n belongs to $(\mathbb{Z} \oplus \mathbb{Z})^+$ for all sufficiently large n.

(iv) Show that $\beta(G^+) = (\mathbb{Z} + \gamma\mathbb{Z}) \cap \mathbb{R}^+$, and conclude that G is order isomorphic to $\mathbb{Z} + \gamma\mathbb{Z}$.

Exercise 6.10. Let $G_1 \xrightarrow{\alpha_1} G_2 \xrightarrow{\alpha_2} G_3 \xrightarrow{\alpha_3} \cdots$ be a sequence of ordered Abelian groups with inductive limit $(G, \{\beta_n\})$, and let $\alpha_{m,n}: G_n \to G_m$ be the associated connecting maps. Let x be an element in G_n. Show that $\beta_n(x)$ belongs to G^+ if and only if there is $m \geqslant n$ such that $\alpha_{m,n}(x)$ belongs to G_m^+. [Hint: If $\beta_n(x)$ belongs to G^+, then $\beta_n(x) = \beta_k(y)$ for some k and some y in G_k^+. Show that $\alpha_{m,n}(x) = \alpha_{m,k}(y)$ for some $m \geqslant \max\{n, k\}$.]

Chapter 7

Classification of AF-Algebras

This chapter contains one of the earliest applications of K-theory to C^*-algebras: the classification of AF-algebras by their ordered K_0-groups — a result obtained by G. Elliott in the early 1970s, see [18].

AF-algebras are inductive limits of finite dimensional C^*-algebras. It is shown in the first section that a C^*-algebra is finite dimensional if and only if it is a direct sum of matrix algebras over $\mathbb{C}$. Some of the basic properties of AF-algebras are described in the second section. Elliott's classification theorem is proved in the third section, and the substantive part of the chapter ends with a section describing and classifying a special class of AF-algebras, the so-called UHF-algebras.

7.1   Finite dimensional C^*-algebras

7.1.1 Matrix units. The matrix algebra $M_n(\mathbb{C})$ and, more generally, the direct sum

$$A = M_{n_1}(\mathbb{C}) \oplus M_{n_2}(\mathbb{C}) \oplus \cdots \oplus M_{n_r}(\mathbb{C})$$

of matrix algebras, where r and $n_1, n_2, \ldots, n_r$ are positive integers, is described by its *standard matrix units*, which are given as follows. Let $e(n, i, j)$ be the (i, j)th standard matrix unit in $M_n(\mathbb{C})$, i.e., the matrix whose (i, j)th entry is 1 and whose other entries are 0. For $1 \leqslant k \leqslant r$ and $1 \leqslant i, j \leqslant n_k$ put

$$e_{ij}^{(k)} = (0, \ldots, 0, e(n_k, i, j), 0, \ldots, 0) \in A,$$

with $e(n_k, i, j)$ in the kth summand. The matrix units satisfy the following identities:

(i) $e_{ij}^{(k)} e_{jl}^{(k)} = e_{il}^{(k)}$,

(ii) $e_{ij}^{(k)} e_{mn}^{(l)} = 0$ if $k \neq l$ or if $j \neq m$,

(iii) $(e_{ij}^{(k)})^* = e_{ji}^{(k)}$,

(iv) $A = \text{span}\{e_{ij}^{(k)} : 1 \leqslant k \leqslant r;\ 1 \leqslant i, j \leqslant n_k\}$.

If B is another C^*-algebra that contains elements $f_{ij}^{(k)}$ for $1 \leqslant k \leqslant r$ and $1 \leqslant i, j \leqslant n_k$ satisfying the analogues of (i), (ii), and (iii) above, with $f_{ij}^{(k)}$ in the place of $e_{ij}^{(k)}$, then there is one and only one *-homomorphism $\varphi \colon A \to B$ such that $\varphi(e_{ij}^{(k)}) = f_{ij}^{(k)}$ for all i, j, k. Indeed, the system of matrix units $\{e_{ij}^{(k)}\}$ is a linear basis for A, and hence there is one and only one *linear map* $\varphi \colon A \to B$ with $\varphi(e_{ij}^{(k)}) = f_{ij}^{(k)}$. One can next use the matrix unit identities (i), (ii), and (iii) to see that φ is a *-homomorphism.

The system $\{f_{ij}^{(k)}\}$ is called a *system of matrix units in B of type A*. If all $f_{ij}^{(k)}$ are non-zero, then φ is injective; and φ is surjective if $\text{span}\{f_{ij}^{(k)}\} = B$. In particular, if the system $\{f_{ij}^{(k)}\}$ is non-zero and satisfies (i), (ii), (iii), and (iv) above, then B is isomorphic to A.

From Example 3.3.2 and Propositions 4.3.4 and 5.1.9 (see also Exercise 5.4), we get that $K_0(A)$ is isomorphic to $\mathbb{Z}^r$ as an ordered Abelian group, where $(\mathbb{Z}^r)^+$ is the set of r-tuples $(x_1, x_2, \ldots, x_r)$ with $x_j \geqslant 0$. More specifically,

$$
\begin{aligned}
K_0(A) &= \ \mathbb{Z}[e_{11}^{(1)}]_0 + \ \mathbb{Z}[e_{11}^{(2)}]_0 + \cdots + \ \mathbb{Z}[e_{11}^{(r)}]_0 \cong \mathbb{Z}^r, \\
\text{(7.1)} \qquad K_0(A)^+ &= \mathbb{Z}^+[e_{11}^{(1)}]_0 + \mathbb{Z}^+[e_{11}^{(2)}]_0 + \cdots + \mathbb{Z}^+[e_{11}^{(r)}]_0 \cong (\mathbb{Z}^+)^r, \\
[1_A]_0 &= \ n_1[e_{11}^{(1)}]_0 + \ n_2[e_{11}^{(2)}]_0 + \cdots + \ n_r[e_{11}^{(r)}]_0.
\end{aligned}
$$

Lemma 7.1.2. *Suppose that* $\{f_{ii}^{(k)} : 1 \leqslant k \leqslant r,\ 1 \leqslant i \leqslant n_k\}$ *is a set of mutually orthogonal projections in a C^*-algebra B, and that*

$$
\text{(7.2)} \qquad\qquad f_{11}^{(k)} \sim f_{22}^{(k)} \sim \cdots \sim f_{n_k n_k}^{(k)}
$$

for $1 \leqslant k \leqslant r$ (where $\sim$ is Murray–von Neumann equivalence). Then there is a system of matrix units $\{f_{ij}^{(k)}\}$ in B that extends the given system $\{f_{ii}^{(k)}\}$.

Proof. By the assumption (7.2) there are partial isometries $f_{1i}^{(k)}$ in B for $1 \leqslant k \leqslant r$ and $1 \leqslant i \leqslant n_k$ such that

$$
f_{1i}^{(k)} f_{1i}^{(k)*} = f_{11}^{(k)}, \qquad f_{1i}^{(k)*} f_{1i}^{(k)} = f_{ii}^{(k)}.
$$

The system

$$f_{ij}^{(k)} = f_{1i}^{(k)^*} f_{1j}^{(k)}$$

in B is then as desired. □

A sub-C^*-algebra D of a C^*-algebra A is called a *maximal Abelian subalgebra* or a *masa* if it is Abelian and not properly contained in any other Abelian sub-C^*-algebra of A. An application of Zorn's lemma shows that every Abelian sub-C^*-algebra is contained in a masa, and hence that every C^*-algebra contains a masa. As a relevant example, the set of diagonal matrices is a masa in $M_n(\mathbb{C})$.

When X is a subset of A, let X' denote the set of all elements a in A for which $ax = xa$ for all elements x in X. It is routine to verify that X' is a norm-closed subalgebra of A, and that X' is a sub-C^*-algebra of A if X is self-adjoint (in the sense that $x^* \in X$ whenever $x \in X$). A sub-C^*-algebra B of A is Abelian if and only if B is contained in B'. A masa is characterized by the following result.

Lemma 7.1.3. *A sub-C^*-algebra D of a C^*-algebra A is a masa if and only if $D' = D$.*

Proof. Suppose that D is a masa. Then D is contained in D', and D' is a C^*-algebra. Since a C^*-algebra is generated by its self-adjoint elements, it will follow that $D = D'$ if each self-adjoint element in D' belongs to D. Let a be a self-adjoint element in D', and consider the set $X = D \cup \{a\}$. Because X is self-adjoint,

$$X \subseteq X' \Rightarrow C^*(X) \subseteq X' \Rightarrow X \subseteq C^*(X)' \Rightarrow C^*(X) \subseteq C^*(X)',$$

which shows that $C^*(X)$ is an Abelian sub-C^*-algebra of A. Hence $C^*(X) = D$ and a belongs to D.

Assume conversely that $D' = D$. Then D is Abelian. Suppose that E is an Abelian sub-C^*-algebra of A that contains D. Then

$$D \subseteq E \subseteq E' \subseteq D' = D,$$

which entails that $E = D$. Therefore D is a masa. □

Lemma 7.1.4. *Let D be a masa in a C^*-algebra A.*

(i) *If a is an element in A that commutes with every element in D, then a belongs to D.*

(ii) *If D is unital, then A is unital, and the unit of D is equal to the unit of A.*

(iii) *If p is a projection in D satisfying $pDp = \mathbb{C}p$, then $pAp = \mathbb{C}p$, i.e., a minimal projection in D is also a minimal projection in A.*

Proof. Part (i) follows immediately from Lemma 7.1.3.

(ii). We show that $a1_D = a$ for every a in A; this will imply that 1_D is a unit for A because then $1_D a = (a^*1_D)^* = a$.

Let a be an element in A, and put $z = a - a1_D$. Then $zd = 0$ for each d in D. Since D is self-adjoint, this implies that also $0 = (zd^*)^* = dz^*$ for all d in D. Hence $d(z^*z) = 0 = (z^*z)d$ for all d in D. This entails that z^*z belongs to D by (i), and $(z^*z)(z^*z) = 0$. We conclude that $\|z\|^4 = \|z^*z\|^2 = \|(z^*z)^2\| = 0$, and that $z = 0$ as desired.

(iii). Let a be an element in pAp, and let d in D be given. Since $pd = dp = pdp = \lambda p$ for some λ in $\mathbb{C}$, we have $ad = \lambda a = da$. By (i), a belongs to D and hence to $pDp = \mathbb{C}p$. □

Proposition 7.1.5. *Any finite dimensional C^*-algebra is isomorphic to*

$$M_{n_1}(\mathbb{C}) \oplus M_{n_2}(\mathbb{C}) \oplus \cdots \oplus M_{n_r}(\mathbb{C})$$

for some positive integers r and $n_1, n_2, \ldots, n_r$.

Proof. Let A be a finite dimensional C^*-algebra, and choose a masa D in A. By Gelfand's theorem (Theorem 1.2.3), D is isomorphic to $C_0(X)$ for some locally compact Hausdorff space X. The set X must be finite, because if it were infinite, then $C_0(X)$ would be infinite dimensional, contradicting that A is finite dimensional. In particular X is compact, D is unital, and hence A is unital by Lemma 7.1.4 (ii). Let $x_1, x_2, \ldots, x_N$ denote the points in X.

Let $p_1, \ldots, p_N$ in D correspond to the functions $q_1, \ldots, q_N$ in $C_0(X) = C(X)$, given by $q_j(x_i) = \delta_{i,j}$. Then $p_1, p_2, \ldots, p_N$ are projections in D,

$$p_1 + p_2 + \cdots + p_N = 1, \quad \text{and} \quad p_j D p_j = \mathbb{C}p_j,$$

because $q_1, q_2, \ldots, q_N$ have these properties in $C(X)$. By Lemma 7.1.4 (iii), $p_j A p_j = \mathbb{C}p_j$ for all j.

Assume that $p_j A p_i$ is non-zero and choose v in $p_j A p_i$ with $\|v\| = 1$. Then v^*v is a positive element of norm equal to 1 in $p_i A p_i = \mathbb{C}p_i$, and consequently $v^*v = p_i$. A similar argument shows that $vv^* = p_j$. If a is any element in

$p_j A p_i$, then $a = a p_i = (a v^*) v$. As $a v^*$ belongs to $p_j A p_j = \mathbb{C} p_j$ and $p_j v = v$ we conclude that a is a scalar multiple of v. Summing up, we have shown that the following alternative holds: either $p_j A p_i = \{0\}$ or $p_j \sim p_i$ in A, and in the latter case, $p_j A p_i = \mathbb{C} v$, where v is a partial isometry satisfying $v^* v = p_i$ and $v v^* = p_j$.

Partition the set $\{p_1, p_2, \ldots, p_N\}$ into Murray–von Neumann equivalence classes. Let r be the number of equivalence classes, and let n_k be the number of elements in the kth equivalence class. Rename the projections p_j in such a way that the elements in the kth equivalence class are $\{f_{11}^{(k)}, f_{22}^{(k)}, \ldots, f_{n_k n_k}^{(k)}\}$. Then

$$\{p_1, p_2, \ldots, p_N\} = \{f_{ii}^{(k)} : 1 \leqslant k \leqslant r, \ 1 \leqslant i \leqslant n_k\},$$

$f_{ii}^{(k)} A f_{jj}^{(l)} = \{0\}$ if $k \neq l$, and $f_{ii}^{(k)} \sim f_{jj}^{(k)}$.

The system $\{f_{ii}^{(k)}\}$ can by Lemma 7.1.2 be extended to a system of matrix units $\{f_{ij}^{(k)}\}$ in A. The argument above shows that $f_{ii}^{(k)} A f_{jj}^{(k)} = \mathbb{C} f_{ij}^{(k)}$.

Let a in A be given. Then, since $1 = \sum_{k,i} f_{ii}^{(k)}$, we have

$$a = \left(\sum_{k,i} f_{ii}^{(k)} \right) a \left(\sum_{k,i} f_{ii}^{(k)} \right) = \sum_{k=1}^{r} \sum_{i,j=1}^{n_k} f_{ii}^{(k)} a f_{jj}^{(k)} = \sum_{k=1}^{r} \sum_{i,j=1}^{n_k} \lambda_{ij}^{(k)} f_{ij}^{(k)}$$

for some $\lambda_{ij}^{(k)}$ in $\mathbb{C}$. This shows that

$$A = \operatorname{span}\{f_{ij}^{(k)} : 1 \leqslant k \leqslant r, \ 1 \leqslant i, j \leqslant n_k\}.$$

It has now been shown that the system $\{f_{ij}^{(k)}\}$ satisfies (i), (ii), (iii), (iv) in Paragraph 7.1.1, and by Paragraph 7.1.1, this completes the proof. □

7.2 AF-algebras

This section contains a brief overview of some fundamental facts about AF-algebras and dimension groups, mostly without proofs. Except for the definition of an AF-algebra given below, these results will not be used in the proof of the classification theorem in Section 7.3, but they are included to give a flavor of the rich theory behind these C^*-algebras and their invariants. Much more can be said.

Definition 7.2.1. An AF-*algebra* is a C^*-algebra which is (isomorphic to) the inductive limit of a sequence of finite dimensional C^*-algebras.

The term "AF" is an abbreviation of Approximately Finite dimensional. It is an automatic consequence of this definition of an AF-algebra that it is separable. It should be mentioned that some expositions allow AF-algebras to be non-separable by defining them to be inductive limits of arbitrary *directed sets* of finite dimensional C^*-algebras. We mention without proof the following local characterization of AF-algebras due to O. Bratteli (see [7]).

Proposition 7.2.2. *A separable C^*-algebra A is an* AF-*algebra if and only if for every $\varepsilon > 0$ and for every finite subset $\{a_1, a_2, \ldots, a_n\}$ of A there exist a finite dimensional sub-C^*-algebra B of A and elements $b_1, b_2, \ldots, b_n$ in B such that $\|a_j - b_j\| < \varepsilon$ for every j.*

7.2.3 Bratteli diagrams. There is a systematic way to write down inductive sequences of finite dimensional C^*-algebras, namely the so-called *Bratteli diagrams*.

For example, let $f_0, f_1, f_2, \ldots$ be the Fibonacci numbers given by $f_0 = f_1 = 1$, and $f_n = f_{n-1} + f_{n-2}$ for $n \geqslant 2$. Put $A_n = M_{f_n}(\mathbb{C}) \oplus M_{f_{n-1}}(\mathbb{C})$, and let $\varphi_n \colon A_n \to A_{n+1}$ be the *-homomorphism given by

$$(x, y) \mapsto \left(\begin{pmatrix} x & 0 \\ 0 & y \end{pmatrix}, x \right).$$

The resulting sequence $A_1 \xrightarrow{\varphi_1} A_2 \xrightarrow{\varphi_2} A_3 \xrightarrow{\varphi_3} \cdots$ of finite dimensional C^*-algebras defines an AF-algebra. (The reader may recognize that the sequence of ordered Abelian groups in Exercise 6.9 is the sequence of ordered K_0-groups arising from this example.) The sequence can be encoded in the following *Bratteli diagram.*

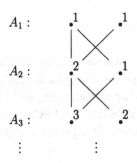

The nth row in the Bratteli diagram corresponds to the algebra A_n. The row consists of a number of *vertices*, each with an attached integer k, representing a direct summand of A_n isomorphic to $M_k(\mathbb{C})$. Between a vertex in row n and a vertex in row $n + 1$ there is a number of *edges* (this number could be 0, 1, 2, etc.) that reflects the properties of the *-homomorphism $\varphi_n\colon A_n \to A_{n+1}$. More specifically, the two vertices represent a direct summand, isomorphic to $M_k(\mathbb{C})$, of A_n and a direct summand, isomorphic to $M_l(\mathbb{C})$, of A_{n+1}. The number of edges between the two vertices is the multiplicity of the *-homomorphism

$$M_k(\mathbb{C}) \xrightarrow{\ \iota\ } A_n \xrightarrow{\ \varphi_n\ } A_{n+1} \xrightarrow{\ \pi\ } M_l(\mathbb{C}).$$

The *multiplicity* of a *-homomorphism $\psi\colon M_k(\mathbb{C}) \to M_l(\mathbb{C})$ is defined to be the number $\mathrm{Tr}(\psi(e))/\mathrm{Tr}(e)$, where e is any non-zero projection in $M_k(\mathbb{C})$, and this number is independent of e. The *-homomorphism $x \mapsto \mathrm{diag}(x,\ldots,x,0)$, with k copies of x, has multiplicity k.

Here are two more Bratteli diagrams.

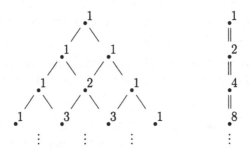

We leave it to the reader to ponder what inductive limits of finite dimensional C^*-algebras they represent.

Definition 7.2.4 (Dimension groups). A *dimension group* is an ordered Abelian group which is (isomorphic to) the inductive limit of the sequence of ordered Abelian groups

$$\mathbb{Z}^{n_1} \xrightarrow{\ \alpha_1\ } \mathbb{Z}^{n_2} \xrightarrow{\ \alpha_2\ } \mathbb{Z}^{n_3} \longrightarrow \cdots$$

for some positive integers n_j and some positive group homomorphisms α_j, where $\mathbb{Z}^n$ is given the usual ordering

$$(\mathbb{Z}^n)^+ = \{(x_1, x_2, \ldots, x_n) \in \mathbb{Z}^n : x_j \geqslant 0\}.$$

Dimension groups can be characterized intrinsically, as described in the theorem below that we state without proof. See the original paper [17], Effros' book [16], or Goodearl's book [22] for a proof. First a definition.

Definition 7.2.5. Let (G, G^+) be an ordered Abelian group, and recall that $x \leqslant y$ if $y - x$ belongs to G^+.

If every x in G for which $nx \geqslant 0$ for some n in $\mathbb{N}$ satisfies $x \geqslant 0$, then G is said to be *unperforated*.

If for every x_1, x_2, y_1, y_2 in G where $x_i \leqslant y_j$ for $i, j = 1, 2$, there exists z in G with $x_i \leqslant z \leqslant y_j$ for $i, j = 1, 2$, then G is said to have the *Riesz interpolation property*.

Theorem 7.2.6 (Effros–Handelman–Shen). *A countable ordered Abelian group (G, G^+) is a dimension group if and only if it is unperforated and has the Riesz interpolation property.*

Example 7.2.7.

(i) Every totally ordered, countable, Abelian group G is a dimension group. Recall that an order relation $\leqslant$ is total if for every pair of elements x, y in G either $x \leqslant y$ or $y \leqslant x$.

In particular, every countable subgroup G of $\mathbb{R}$ equipped with the ordering from $\mathbb{R}$, viz. $G^+ = G \cap \mathbb{R}^+$, is totally ordered and hence a dimension group.

(ii) An ordered Abelian group (G, G^+) is called a *lattice* if each pair of elements in G has a least upper bound and a greatest lower bound. Every countable unperforated ordered Abelian group which is a lattice is a dimension group.

(iii) The direct sum of two dimension groups is a dimension group: Let G and H be dimension groups, set $D = G \oplus H$, and set

$$D^+ = \{(g, h) : g \geqslant 0, h \geqslant 0\}.$$

Then (D, D^+) is a dimension group.

(iv) Let G and H be ordered Abelian groups, set $D = G \oplus H$, and set

$$D^+ = \{(g, h) : g > 0, h > 0\} \cup \{(0, 0)\}.$$

Then (D, D^+) is an ordered Abelian group. However, (D, D^+) is usually not a dimension group, even if both G and H are dimension groups.

For example, if $G = \mathbb{Z}$ with the usual order and if H contains elements h_1, h_2, h_1', h_2' such that $h_1 \neq h_2$, $h_1' \neq h_2'$ and $h_i < h_j'$ for all i, j, then D is not a dimension group. (There is no interpolating element for $x_1 = (0, h_1)$, $x_2 = (0, h_2)$, $y_1 = (1, h_1')$, $y_2 = (1, h_2')$.)

On the other hand, if $G = H = \mathbb{Q}$ equipped with the standard order, then D is a dimension group.

(v) Let G be a dimension group, let H be any countable, torsion free, Abelian group, set $D = G \oplus H$, and set

$$D^+ = \{(g, h) : g > 0\} \cup \{(0, 0)\}.$$

Then (D, D^+) is an ordered Abelian group. If $G = \mathbb{Z}$ and $H \neq \{0\}$, then D is not a dimension group. (There is no interpolating element for $x_1 = (0, 0)$, $x_2 = (0, h)$, $y_1 = (1, 0)$, $y_2 = (1, h)$ if h is a non-zero element in H.)

If G is a dense subgroup of $\mathbb{R}$, then D is a dimension group.

(vi) The ordered Abelian group in Exercise 5.5 is not a dimension group because it is not unperforated.

Every AF-algebra is stably finite (see Exercise 7.5), and $(K_0(A), K_0(A)^+)$ is therefore an ordered Abelian group for every unital AF-algebra A by Proposition 5.1.5. One can also use continuity of K_0 (Theorem 6.3.2) to reach this conclusion — actually for all AF-algebras, unital or not.

The proof of the proposition below uses a lemma from the next section.

Proposition 7.2.8. *The ordered K_0-group $(K_0(A), K_0(A)^+)$ of an AF-algebra A is a dimension group. And conversely, every dimension group is order isomorphic to the ordered K_0-group of some AF-algebra.*

Proof. Let A be an AF-algebra, and let $A_1 \to A_2 \to A_3 \to \cdots$ be a sequence of finite dimensional C^*-algebras with inductive limit A. By continuity of K_0 (Theorem 6.3.2), $K_0(A)$ is isomorphic to the inductive limit of the sequence $K_0(A_1) \to K_0(A_2) \to K_0(A_3) \to \cdots$ in the category of ordered Abelian groups. As shown in Paragraph 7.1.1, each $K_0(A_n)$ is order isomorphic to $\mathbb{Z}^{k_n}$ with its standard order for some k_n in $\mathbb{N}$, and so we conclude that $K_0(A)$ is a dimension group.

Suppose conversely that (G, G^+) is a dimension group. Then G is isomorphic to the inductive limit of the sequence

$$\mathbb{Z}^{n_1} \xrightarrow{\alpha_1} \mathbb{Z}^{n_2} \xrightarrow{\alpha_2} \mathbb{Z}^{n_3} \xrightarrow{\alpha_3} \cdots$$

for suitable n_j and suitable positive group homomorphisms α_j. Choose inductively order units

$$u_j = (k_1^{(j)}, k_2^{(j)}, \ldots, k_{n_j}^{(j)}) \in (\mathbb{Z}^{n_j})^+$$

such that $\alpha_j(u_j) \leqslant u_{j+1}$ for each natural number j. Because u_j is an order unit, we have $k_i^{(j)} \geqslant 1$ for all i, j. For each natural number j put

$$A_j = M_{k_1^{(j)}}(\mathbb{C}) \oplus M_{k_2^{(j)}}(\mathbb{C}) \oplus \cdots \oplus M_{k_{n_j}^{(j)}}(\mathbb{C}),$$

and let $\gamma_j \colon \mathbb{Z}^{n_j} \to K_0(A_j)$ be the canonical order isomorphism from equation (7.1) satisfying $\gamma_j(u_j) = [1_{A_j}]_0$. Use Lemma 7.3.2 below to find a *-homomorphism $\varphi_j \colon A_j \to A_{j+1}$ for each j such that the diagram

$$
\begin{array}{ccccccc}
\mathbb{Z}^{n_1} & \xrightarrow{\alpha_1} & \mathbb{Z}^{n_2} & \xrightarrow{\alpha_2} & \mathbb{Z}^{n_3} & \longrightarrow & \cdots \\
\gamma_1 \downarrow & & \gamma_2 \downarrow & & \gamma_3 \downarrow & & \\
K_0(A_1) & \xrightarrow[K_0(\varphi_1)]{} & K_0(A_2) & \xrightarrow[K_0(\varphi_2)]{} & K_0(A_3) & \longrightarrow & \cdots
\end{array}
$$

is commutative. Let A be the inductive limit of the sequence

$$A_1 \xrightarrow{\varphi_1} A_2 \xrightarrow{\varphi_2} A_3 \xrightarrow{\varphi_3} \cdots .$$

Then A is an AF-algebra, and by continuity of K_0 (Theorem 6.3.2), $K_0(A)$ is order isomorphic to G. $\qquad\square$

7.3   Elliott's classification theorem

This section contains Elliott's classification of (unital) AF-algebras. We begin with a discussion of the cancellation property of C^*-algebras and two preliminary lemmas of some independent interest.

Definition 7.3.1. A C^*-algebra A is said to have the *cancellation property* if the semigroup $\mathcal{D}(A)$ has the cancellation property, see Paragraph 3.1.1 and Definition 2.3.3. Equivalently, A has the cancellation property if and only if, for every pair of projections p, q in $\mathcal{P}_\infty(A)$,

$$[p]_0 = [q]_0 \iff p \sim_0 q.$$

It follows from Example 3.3.2 and Exercise 2.9 that $M_n(\mathbb{C})$ has the cancellation property. Using this, Exercise 7.4, and Proposition 7.1.5, it follows that every finite dimensional C^*-algebra has the cancellation property. This is all we shall need about the cancellation property in this section.

Every AF-algebra has the cancellation property by Exercise 7.4. There is a more general result in this direction due to M. Rieffel: a C^*-algebra A with the property that $\mathrm{GL}(\widetilde{A})$ is a dense subset of $\widetilde{A}$ is said to have *stable rank one*, in symbols, $\mathrm{sr}(A) = 1$. Every C^*-algebra of stable rank one has the cancellation property by [37].

Recall (from Exercise 4.10) that if u is a unitary element in a unital C^*-algebra B, then $\mathrm{Ad}\, u$ is the automorphism on B given by $(\mathrm{Ad}\, u)(b) = ubu^*$.

Lemma 7.3.2. *Let A be a finite dimensional C^*-algebra, and let B be a unital C^*-algebra with the cancellation property.*

(i) *For each positive group homomorphism $\alpha \colon K_0(A) \to K_0(B)$ satisfying $\alpha([1_A]_0) \leqslant [1_B]_0$ there is a $*$-homomorphism $\varphi \colon A \to B$ with $K_0(\varphi) = \alpha$.*

    If $\alpha([1_A]_0) = [1_B]_0$, then φ is necessarily unit preserving.

(ii) *Let $\varphi, \psi \colon A \to B$ be $*$-homomorphisms. Then $K_0(\varphi) = K_0(\psi)$ if and only if $\psi = \mathrm{Ad}\, u \circ \varphi$ for some unitary u in B.*

Proof. (i). We show first that if $g_1, g_2, \ldots, g_N$ are elements in $K_0(B)^+$ satisfying $\sum g_j \leqslant [1_B]_0$, then there are mutually orthogonal projections $p_1, p_2, \ldots, p_N$ in B with $[p_j]_0 = g_j$. For this it suffices to show that if p in $\mathcal{P}(B)$ and g in $K_0(B)$ are such that $0 \leqslant g \leqslant [1_B]_0 - [p]_0$, then there is a projection q in B which is orthogonal to p and which satisfies $g = [q]_0$. To prove this claim take e in $\mathcal{P}_n(B)$ and f in $\mathcal{P}_m(B)$ such that $[e]_0 = g$ and $[f]_0 = [1_B]_0 - [p]_0 - g$. Using that B has the cancellation property, we derive that $e \oplus f \sim_0 1_B - p$, and so $e \oplus f = v^*v$ and $1_B - p = vv^*$ for some v in $M_{1,n+m}(B)$. Put $q = v(e \oplus 0_m)v^*$. Then q is a projection in B, $q \leqslant 1_B - p$, and $q \sim_0 e$, whence $[q]_0 = [e]_0 = g$.

Let $\{e_{ij}^{(k)}\}$, $1 \leqslant k \leqslant r$, $1 \leqslant i,j \leqslant n_k$, be the system of standard matrix units for A. By the result obtained above, there are pairwise orthogonal projections $f_{ii}^{(k)}$, $1 \leqslant k \leqslant r$, $1 \leqslant i \leqslant n_k$, in B with $\alpha([e_{ii}^{(k)}]_0) = [f_{ii}^{(k)}]_0$.

Using once again that B has the cancellation property, we obtain

$$e_{ii}^{(k)} \sim e_{jj}^{(k)} \;\Rightarrow\; [e_{ii}^{(k)}]_0 = [e_{jj}^{(k)}]_0 \;\Rightarrow\; [f_{ii}^{(k)}]_0 = [f_{jj}^{(k)}]_0 \;\Rightarrow\; f_{ii}^{(k)} \sim f_{jj}^{(k)}.$$

By Lemma 7.1.2 the system $\{f_{ii}^{(k)}\}$ has an extension to a system of matrix units $\{f_{ij}^{(k)}\}$ in B. Paragraph 7.1.1 produces a $*$-homomorphism $\varphi \colon A \to B$

such that $\varphi(e_{ij}^{(k)}) = f_{ij}^{(k)}$.

As remarked in Paragraph 7.1.1, the elements $[e_{11}^{(1)}]_0, [e_{11}^{(2)}]_0, \ldots, [e_{11}^{(r)}]_0$ generate $K_0(A)$,

$$K_0(\varphi)([e_{11}^{(k)}]_0) = [\varphi(e_{11}^{(k)})]_0 = [f_{11}^{(k)}]_0 = \alpha([e_{11}^{(k)}]_0),$$

and so $K_0(\varphi) = \alpha$.

Assume that $\alpha([1_A]_0) = [1_B]_0$, and put $p = \sum_{i,k} f_{ii}^{(k)}$. Then p is a projection in B, and $\varphi(1_A) = p$. Consequently

$$[1_B - p]_0 = [1_B]_0 - [p]_0 = \alpha([1_A]_0) - K_0(\varphi)([1_A]_0) = 0.$$

Hence $1_B - p \sim_0 0$ because B has the cancellation property. This entails that $1_B - p = 0$, and so $\varphi(1_A) = p = 1_B$.

(ii). Suppose that $K_0(\varphi) = K_0(\psi)$. Retain the matrix units $\{e_{ij}^{(k)}\}$ for A. Then

$$[\varphi(e_{11}^{(k)})]_0 = K_0(\varphi)([e_{11}^{(k)}]_0) = K_0(\psi)([e_{11}^{(k)}]_0) = [\psi(e_{11}^{(k)})]_0,$$
$$[1_B - \varphi(1_A)]_0 = [1_B]_0 - K_0(\varphi)([1_A]_0) = [1_B]_0 - K_0(\psi)([1_A]_0)$$
$$= [1_B - \psi(1_A)]_0.$$

Because B has the cancellation property, there are partial isometries $v_1, v_2, \ldots,$ v_r and w in B such that

$$v_k^* v_k = \varphi(e_{11}^{(k)}), \qquad v_k v_k^* = \psi(e_{11}^{(k)}), \quad 1 \leqslant k \leqslant r,$$
$$w^* w = 1_B - \varphi(1_A), \qquad w w^* = 1_B - \psi(1_A).$$

Use Exercise 2.6 and the fact that $\psi(e_{i1}^{(k)}) v_k \varphi(e_{1i}^{(k)})$ is a partial isometry with range projection $\psi(e_{ii}^{(k)})$ and support projection $\varphi(e_{ii}^{(k)})$ to see that

$$u = w + \sum_{k=1}^r \sum_{i=1}^{n_k} \psi(e_{i1}^{(k)}) v_k \varphi(e_{1i}^{(k)})$$

is unitary. Moreover

$$u\varphi(e_{ij}^{(k)}) = \psi(e_{i1}^{(k)}) v_k \varphi(e_{1j}^{(k)}) = \psi(e_{ij}^{(k)}) u,$$

which shows that $u\varphi(e_{ij}^{(k)})u^* = \psi(e_{ij}^{(k)})$ for all i, j, k. It follows that $\operatorname{Ad} u \circ \varphi = \psi$ as desired.

If $\psi = \mathrm{Ad}\, u \circ \varphi$, then

$$K_0(\psi) = K_0(\mathrm{Ad}\, u \circ \varphi) = K_0(\mathrm{Ad}\, u) \circ K_0(\varphi) = K_0(\varphi),$$

see Exercise 4.10. $\square$

Lemma 7.3.3. *Let*

$$A_1 \xrightarrow{\varphi_1} A_2 \xrightarrow{\varphi_2} A_3 \xrightarrow{\varphi_3} \cdots$$

be a sequence of finite dimensional C^*-algebras with inductive limit $(A, \{\mu_n\})$. Let B be a finite dimensional C^*-algebra, and assume that there are positive group homomorphisms $\alpha\colon K_0(A_1) \to K_0(B)$ and $\gamma\colon K_0(B) \to K_0(A)$ such that $\gamma \circ \alpha = K_0(\mu_1)$.

Then, for some natural number n, there is a positive group homomorphism $\beta\colon K_0(B) \to K_0(A_n)$ making the diagram

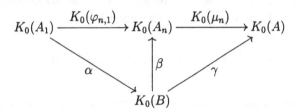

commutative. If all the connecting maps $\varphi_1, \varphi_2, \ldots$ are unit preserving and if $\alpha([1_{A_1}]_0) = [1_B]_0$, then $\beta([1_B]_0) = [1_{A_n}]_0$.

Proof. Let $\{e_{ij}^{(k)}\}$, $1 \leqslant k \leqslant r$, $1 \leqslant i, j \leqslant n_k$, be the matrix units for B, and put $x_k = \gamma([e_{11}^{(k)}]_0)$ in $K_0(A)^+$. From continuity of K_0 (Theorem 6.3.2) we have

$$K_0(A)^+ = \bigcup_{m=1}^{\infty} K_0(\mu_m)\,(K_0(A_m)^+),$$

and so we can find a natural number m and $y_1, \ldots, y_r$ in $K_0(A_m)^+$ such that $x_k = K_0(\mu_m)(y_k)$ for all k. From the description of the K_0-group of a finite dimensional C^*-algebra in Paragraph 7.1.1 we see that $K_0(B)$ is the free Abelian group generated by $[e_{11}^{(1)}]_0, [e_{11}^{(2)}]_0, \ldots, [e_{11}^{(r)}]_0$, and hence there is a (unique) group homomorphism $\beta'\colon K_0(B) \to K_0(A_m)$ with

$$\beta'([e_{11}^{(1)}]_0) = y_1, \quad \beta'([e_{11}^{(2)}]_0) = y_2, \quad \ldots, \quad \beta'([e_{11}^{(r)}]_0) = y_r.$$

The description in Paragraph 7.1.1 also shows that if g is a positive element in $K_0(B)$, then

$$g = m_1[e_{11}^{(1)}]_0 + m_2[e_{11}^{(2)}]_0 + \cdots + m_r[e_{11}^{(r)}]_0$$

for some non-negative integers $m_1, m_2, \ldots, m_r$. It follows that

$$\beta'(g) = m_1 y_1 + m_2 y_2 + \cdots + m_r y_r \in K_0(B)^+.$$

This shows that β' is positive. Since

$$(K_0(\mu_m) \circ \beta')([e_{11}^{(k)}]_0) = K_0(\mu_m)(y_k) = x_k = \gamma([e_{11}^{(k)}]_0),$$

we see that $K_0(\mu_m) \circ \beta' = \gamma$. Notice also that

(7.3) $$K_0(\mu_m) \circ (\beta' \circ \alpha - K_0(\varphi_{m,1})) = \gamma \circ \alpha - K_0(\mu_1) = 0.$$

Let $\{g_1, g_2, \ldots, g_s\}$ be a set of generators of the finitely generated Abelian group $K_0(A_1)$. By continuity of K_0 (Theorem 6.3.2),

$$\operatorname{Ker}(K_0(\mu_m)) = \bigcup_{n=m+1}^{\infty} \operatorname{Ker}(K_0(\varphi_{n,m})),$$

and so $(\beta' \circ \alpha - K_0(\varphi_{m,1}))(g_i)$ all belong to the kernel of $K_0(\varphi_{n,m})$ for some integer $n > m$ by (7.3). Put $\beta = K_0(\varphi_{n,m}) \circ \beta'$. Then $(\beta \circ \alpha - K_0(\varphi_{n,1}))(g_i) = 0$ for all i, and this shows that $\beta \circ \alpha = K_0(\varphi_{n,1})$. Also,

$$\gamma = K_0(\mu_m) \circ \beta' = K_0(\mu_n) \circ K_0(\varphi_{n,m}) \circ \beta' = K_0(\mu_n) \circ \beta.$$

The last claim about the units follows immediately from the commutativity of the diagram.   □

Theorem 7.3.4 (Elliott). *Two unital AF-algebras A and B are isomorphic if and only if the triples $(K_0(A), K_0(A)^+, [1_A]_0)$ and $(K_0(B), K_0(B)^+, [1_B]_0)$ are isomorphic, in other words, if and only if there is a group isomorphism $\alpha \colon K_0(A) \to K_0(B)$ that satisfies $\alpha(K_0(A)^+) = K_0(B)^+$ and $\alpha([1_A]_0) = [1_B]_0$.*

Moreover, for any such group isomorphism α there is a *-isomorphism $\varphi \colon A \to B$ with $K_0(\varphi) = \alpha$.

Proof. By Paragraph 5.1.8, any *-isomorphism $\varphi \colon A \to B$ induces an isomorphism between the triples $(K_0(A), K_0(A)^+, [1_A]_0)$ and $(K_0(B), K_0(B)^+, [1_B]_0)$.

To prove the theorem it therefore suffices to show that every isomorphism α, as in the theorem, is induced by a *-isomorphism $\varphi\colon A \to B$.

Use Exercise 6.7 to find sequences of finite dimensional C^*-algebras

$$A_1 \xrightarrow{f_1} A_2 \xrightarrow{f_2} A_3 \longrightarrow \cdots, \qquad B_1 \xrightarrow{g_1} B_2 \xrightarrow{g_2} B_3 \longrightarrow \cdots,$$

with *unit preserving* connecting maps, and *-homomorphisms $\mu_n\colon A_n \to A$ and $\lambda_n\colon B_n \to B$ such that $(A, \{\mu_n\})$ and $(B, \{\lambda_n\})$ are the inductive limits of these two sequences. Put $B_0 = \mathbb{C}$, and let $g_0\colon B_0 \to B_1$ and $\psi_0\colon B_0 \to A_1$ be the (unique) unital *-homomorphisms. Put $\beta_0 = K_0(\psi_0)$. Use Lemma 7.3.3 inductively to construct the commutative diagram

$$
\begin{array}{ccccccccc}
K_0(A_1) & \longrightarrow & K_0(A_{n_2}) & \longrightarrow & K_0(A_{n_3}) & \to \cdots \to & K_0(A) \\
\nearrow \; \searrow^{\alpha_1} & & \nearrow \; \searrow^{\alpha_2} & & \nearrow & & \alpha^{-1} \uparrow \downarrow \alpha \\
\beta_0 \qquad & & \beta_1 \qquad & & \beta_2 & & \\
K_0(B_0) & \longrightarrow & K_0(B_{m_1}) & \longrightarrow & K_0(B_{m_2}) & \longrightarrow \cdots \longrightarrow & K_0(B)
\end{array}
$$

where $1 = n_1, n_2, n_3, \ldots$ and $0, m_1, m_2, \ldots$ are strictly increasing sequences of integers, and where α_j and β_j are positive group homomorphisms that preserve the classes of the units. To do so, find first m_1 and α_1, using Lemma 7.3.3 on the two maps $\beta_0\colon K_0(B_0) \to K_0(A_1)$ and $\alpha \circ K_0(\mu_1)\colon K_0(A_1) \to K_0(B)$. Continue in a similar way to find n_2, β_1, and m_2, α_2, etc.

The inductive limits of the two sequences

$$A_{n_1} \xrightarrow{f_{n_2, n_1}} A_{n_2} \xrightarrow{f_{n_3, n_2}} A_{n_3} \longrightarrow \cdots,$$

$$B_{m_1} \xrightarrow{g_{m_2, m_1}} B_{m_2} \xrightarrow{g_{m_3, m_2}} B_{m_3} \longrightarrow \cdots$$

are A and B, respectively, by Exercise 6.7 (i). To simplify the notation, we may assume that $n_j = m_j = j$, so that $f_{n_{j+1}, n_j} = f_j$ and $g_{n_{j+1}, n_j} = g_j$.

Use Lemma 7.3.2 (i) to find unital *-homomorphisms $\varphi_j'\colon A_j \to B_j$ and $\psi_j'\colon B_j \to A_{j+1}$ with $K_0(\varphi_j') = \alpha_j$ and $K_0(\psi_j') = \beta_j$ for each j. Because

$$K_0(f_j) = \beta_j \circ \alpha_j = K_0(\psi_j' \circ \varphi_j'),$$
$$K_0(g_j) = \alpha_{j+1} \circ \beta_j = K_0(\varphi_{j+1}' \circ \psi_j'),$$

Lemma 7.3.2 (ii) shows that we can inductively find unitary elements u_j in

A_{j+1} and v_j in B_j with $v_1 = 1_B$ and

$$f_j = \operatorname{Ad} u_j \circ \psi'_j \circ \varphi_j = \psi_j \circ \varphi_j,$$
$$g_j = \operatorname{Ad} v_{j+1} \circ \varphi'_{j+1} \circ \psi_j = \varphi_{j+1} \circ \psi_j,$$

where $\varphi_j = \operatorname{Ad} v_j \circ \varphi'_j$ and $\psi_j = \operatorname{Ad} u_j \circ \psi'_j$. Notice that $K_0(\varphi_j) = K_0(\varphi'_j)$ and $K_0(\psi_j) = K_0(\psi'_j)$ by Lemma 7.3.2 (ii).

We have now arrived at the commutative diagram

$$
\begin{array}{ccccccc}
A_1 & \xrightarrow{f_1} & A_2 & \xrightarrow{f_2} & A_3 \rightarrow \cdots \rightarrow & A \\
& \varphi_1 & & \varphi_2 & & \varphi \,\|\, \psi = \varphi^{-1} \\
B_0 & \xrightarrow{g_0} & B_1 & \xrightarrow{g_1} & B_2 \longrightarrow \cdots \rightarrow & B.
\end{array}
$$

The isomorphisms φ and ψ indicated by the dashed arrows in the diagram exist by Exercise 6.8. To complete the proof we must show that $K_0(\varphi) = \alpha$.

The two diagrams

$$
\begin{array}{ccc}
K_0(A_j) & \xrightarrow{K_0(\mu_j)} & K_0(A) \\
K_0(\varphi_j) = \alpha_j \Big\downarrow & & \Big\downarrow K_0(\varphi) \\
K_0(B_j) & \xrightarrow{K_0(\lambda_j)} & K_0(B)
\end{array}
\qquad
\begin{array}{ccc}
K_0(A_j) & \xrightarrow{K_0(\mu_j)} & K_0(A) \\
\alpha_j \Big\downarrow & & \Big\downarrow \alpha \\
K_0(B_j) & \xrightarrow{K_0(\lambda_j)} & K_0(B)
\end{array}
$$

commute for every j. Hence $K_0(\varphi)$ and α agree on the image of $K_0(\mu_j)$ for every j. By continuity of K_0 (Theorem 6.3.2),

$$K_0(A) = \bigcup_{j=1}^{\infty} \operatorname{Im}(K_0(\mu_j)),$$

and this shows that $K_0(\varphi) = \alpha$ as desired. $\square$

Elliott's classification theorem holds also for non-unital AF-algebras. Obviously, one must use a different invariant from the one used in our (unital) version. One should replace the positive cone of $K_0(A)$ and the position of the unit in $K_0(A)$ with the so-called *dimension range*

$$\mathcal{D}_0(A) = \{[p]_0 : p \,\epsilon\, \mathcal{P}(A)\} \subseteq K_0(A)^+$$

associated with each C^*-algebra A. Two AF-algebras A and B are isomorphic if and only if their invariants, $(K_0(A), \mathcal{D}_0(A))$ and $(K_0(B), \mathcal{D}_0(B))$, are isomorphic, i.e., if and only if there is a group isomorphism $\alpha \colon K_0(A) \to K_0(B)$ that satisfies $\alpha(\mathcal{D}_0(A)) = \mathcal{D}_0(B)$. One can even replace the invariant $(K_0(A), \mathcal{D}_0(A))$ with the single set $\mathcal{D}_0(A)$ equipped with a partially defined addition, whereby two elements in $\mathcal{D}_0(A)$ can be added together if their sum belongs to $\mathcal{D}_0(A)$.

7.4* UHF-algebras

UHF-algebras form a class of particularly accessible AF-algebras. They were studied and classified by Glimm in [21] before general AF-algebras were considered. We shall here derive the classification of UHF-algebras as a consequence of Theorem 7.3.4.

Definition 7.4.1. A UHF-*algebra* ("Uniformly Hyper-Finite C^*-algebra") is a C^*-algebra which is (isomorphic to) the inductive limit of the sequence

$$(7.4) \qquad M_{k_1}(\mathbb{C}) \xrightarrow{\varphi_1} M_{k_2}(\mathbb{C}) \xrightarrow{\varphi_2} M_{k_3}(\mathbb{C}) \xrightarrow{\varphi_3} \cdots$$

for some natural numbers $k_1, k_2, k_3, \ldots$ and some *unit preserving* connecting *-homomorphisms $\varphi_1, \varphi_2, \varphi_3, \ldots$.

There is a unital *-homomorphism $\varphi \colon M_n(\mathbb{C}) \to M_m(\mathbb{C})$ for a given pair of natural numbers n and m, if and only if n divides m, i.e., $m = dn$ for some integer d. The integer d can be recovered from φ by the formula $d = \mathrm{Tr}(\varphi(e))$, where e is any one-dimensional projection in $M_n(\mathbb{C})$. Conversely, if $m = dn$, then there we can take $\varphi \colon M_n(\mathbb{C}) \to M_m(\mathbb{C})$ to be $\varphi(a) = \mathrm{diag}(a, a, \ldots, a)$ (with d copies of a in the diagonal). It follows that we can realize the sequence in (7.4) with unital *-homomorphisms $\varphi_1, \varphi_2, \ldots$ if and only if k_j divides k_{j+1} for every j.

Definition 7.4.2. Let $\{p_1, p_2, \ldots\}$ be the set of all positive prime numbers listed in increasing order. A *supernatural number* is a sequence $n = \{n_j\}_{j=1}^{\infty}$ where each n_j belongs to $\{0, 1, 2, \ldots, \infty\}$. More suggestively, we view n as a formal infinite prime factorization, where n_j is the power of the prime p_j, i.e.,

$$n = \prod_{j=1}^{\infty} p_j^{n_j}.$$

Associate to the supernatural number $n = \{n_j\}_{j=1}^{\infty}$ the subgroup $Q(n)$ of the additive group $\mathbb{Q}$ consisting of all fractions x/y, where x is any integer, and where $y = \prod_{j=1}^{\infty} p_j^{m_j}$ for some positive integers $m_j \leqslant n_j$ (where $m_j > 0$ for only finitely many j).

The product of two supernatural numbers $n = \{n_j\}_{j=1}^{\infty}$ and $m = \{m_j\}_{j=1}^{\infty}$ is defined to be $nm = \{n_j + m_j\}_{j=1}^{\infty}$. Each natural number is in particular a supernatural number, and hence we can multiply a supernatural number by a natural number.

Proposition 7.4.3.

(i) *Every subgroup G of $(\mathbb{Q}, +)$ that contains 1 is equal to $Q(n)$ for some supernatural number n.*

(ii) *Given supernatural numbers n and n', $(Q(n), 1) \cong (Q(n'), 1)$, i.e., there is an isomorphism $\alpha \colon Q(n) \to Q(n')$ with $\alpha(1) = 1$, if and only if $n = n'$.*

(iii) *The groups $Q(n)$ and $Q(n')$ are isomorphic if and only if there are integers m, m' such that $mn = m'n'$.*

Proof. (i). For each j put

$$(7.5) \qquad n_j = \sup\{m \in \mathbb{Z}^+ : p_j^{-m} \in G\} \in \{0, 1, 2, \ldots, \infty\},$$

and let n be the supernatural number $\{n_j\}_{j=1}^{\infty}$. Let $t = x/y$ be a non-zero rational number, where x, y are relatively prime, and write

$$y = p_1^{m_1} p_2^{m_2} \cdots p_k^{m_k}, \qquad m_j \in \mathbb{Z}^+.$$

Put $y_j = y p_j^{-m_j}$. Choose integers a, b such that $1 = ax + by$. Then

$$(7.6) \qquad \frac{1}{p_j^{m_j}} = \frac{y_j}{y} = \frac{y_j(ax + by)}{y} = y_j(at + b).$$

Let $c_1, c_2, \ldots, c_k$ be integers such that $1 = c_1 y_1 + \cdots + c_k y_k$. Then

$$(7.7) \qquad t = \frac{x(c_1 y_1 + c_2 y_2 + \cdots + c_k y_k)}{y} = \frac{x c_1}{p_1^{m_1}} + \frac{x c_2}{p_2^{m_2}} + \cdots + \frac{x c_k}{p_k^{m_k}}.$$

We conclude from (7.6) and (7.7) that

$$t \in G \iff 1/p_1^{m_1}, 1/p_2^{m_2}, \ldots, 1/p_k^{m_k} \in G$$
$$\iff m_1 \leqslant n_1, m_2 \leqslant n_2, \ldots, m_k \leqslant n_k \iff t \in Q(n),$$

and this proves that $G = Q(n)$.

(ii). Suppose that $\alpha \colon Q(n) \to Q(n')$ is an isomorphism with $\alpha(1) = 1$. Then $\alpha(t) = t$ for every t in $Q(n)$, whence $Q(n) = Q(n')$ and $\alpha = \mathrm{id}$.

Conversely, the supernatural number n can be recovered from the group $Q(n)$ by (7.5), and this shows that $n = n'$ if $Q(n) = Q(n')$.

(iii). Let $\alpha \colon Q(n) \to Q(n')$ be an isomorphism. By replacing α by $-\alpha$ if necessary we may assume that $\alpha(1)$ is positive. In that case there are positive integers m, m' with $\alpha(m) = m'$. The composition of the three isomorphisms

$$Q(nm) \xrightarrow{\;m\;} Q(n) \xrightarrow{\;\alpha\;} Q(n') \xrightarrow{\;(m')^{-1}\;} Q(n'm'),$$

where the isomorphisms m and $(m')^{-1}$ are given by $t \mapsto mt$ and $t \mapsto t/m'$, respectively, defines an isomorphism $\beta \colon Q(nm) \to Q(n'm')$ with $\beta(1) = 1$. Hence $nm = n'm'$ by (ii).

Conversely, if $nm = n'm'$ for some natural numbers m, m', then

$$Q(n) \xrightarrow{\;m^{-1}\;} Q(nm) \xrightarrow{\;\mathrm{id}\;} Q(n'm') \xrightarrow{\;m'\;} Q(n')$$

defines an isomorphism from $Q(n)$ onto $Q(n')$. $\qquad\square$

Lemma 7.4.4. *Let A be a UHF-algebra which is the inductive limit of the sequence*

$$(7.8) \qquad M_{k_1}(\mathbb{C}) \xrightarrow{\;\varphi_1\;} M_{k_2}(\mathbb{C}) \xrightarrow{\;\varphi_2\;} M_{k_3}(\mathbb{C}) \longrightarrow \cdots$$

where the connecting maps are unital and where $\{k_j\}_{j=1}^{\infty}$ is a sequence of positive integers that, as they must, satisfy $k_j | k_{j+1}$ for all j. Write

$$k_i = \prod_{j=1}^{\infty} p_j^{n_{i,j}}, \qquad n_{i,j} \in \mathbb{Z}^+,$$

and let n be the supernatural number $\{n_j\}_{j=1}^{\infty}$, where $n_j = \sup\{n_{i,j} : i \in \mathbb{N}\}$.

(i) $Q(n) = \bigcup_{i=1}^{\infty} k_i^{-1} \mathbb{Z}$.

(ii) $(K_0(A), [1]_0) \cong (Q(n), 1)$.

The number n is uniquely determined from A by (ii), see Proposition 7.4.3. We call n the supernatural number associated with the UHF-algebra A.

Proof of lemma. (i). The number k_i^{-1} belongs to $Q(n)$ because $n_{i,j} \leqslant n_j$. It follows that $\bigcup_{i=1}^{\infty} k_i^{-1}\mathbb{Z}$ is contained in $Q(n)$.

Conversely, let $t = x/y$ be an element in $Q(n)$, where x, y are relatively prime and

$$y = p_1^{m_1} p_2^{m_2} \cdots p_r^{m_r}.$$

Then $m_j \leqslant n_j$, and, moreover, there exists i such that $m_j \leqslant n_{i,j}$ for all j. Hence y divides k_i, and so t belongs to $k_i^{-1}\mathbb{Z}$.

(ii). Let τ_j be the *normalized trace* on $M_{k_j}(\mathbb{C})$ that is given by $\tau_j(a) = k_j^{-1}\mathrm{Tr}(a)$, where Tr is the standard trace on $M_{k_j}(\mathbb{C})$. Then $K_0(\tau_j)$ is an isomorphism from $K_0(M_{k_j}(\mathbb{C}))$ onto $k_j^{-1}\mathbb{Z}$, see Example 3.3.2. Since $\tau_{j+1} \circ \varphi_j = \tau_j$, we have $K_0(\tau_{j+1}) \circ K_0(\varphi_j) = K_0(\tau_j)$ for all j. This gives the homomorphism β in the commutative diagram

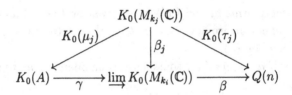

where $\mu_j \colon M_{k_j}(\mathbb{C}) \to A$ and β_j are the inductive limit morphisms, and where γ is the isomorphism from Theorem 6.3.2 (continuity of K_0). Surjectivity of β follows from (i), and β is injective because each $K_0(\tau_j)$ is injective. It follows that $\beta \circ \gamma$ is an isomorphism, and $(\beta \circ \gamma)([1_A]_0) = K_0(\tau_1)([1_{k_1}]_0) = 1$. $\square$

The result below summarizes the classification theorem for UHF-algebras.

Theorem 7.4.5.

(i) *Let A and A' be UHF-algebras with associated supernatural numbers n and n', respectively. The following four conditions are equivalent:*

(a) $A \cong A'$,

(b) $n = n'$,

(c) $(K_0(A), [1]_0) \cong (K_0(A'), [1]_0)$,

(d) $(K_0(A), K_0(A)^+, [1]_0) \cong (K_0(A'), K_0(A')^+, [1]_0)$.

(ii) *For every subgroup G of $(\mathbb{Q}, +)$ containing 1 there is a UHF-algebra A such that $(G, 1)$ is isomorphic to $(K_0(A), [1]_0)$.*

(iii) *For every supernatural number n there is a UHF-algebra whose associated supernatural number is n.*

Proof. (i). Lemma 7.4.4 (ii) and Proposition 7.4.3 (ii) give that (b) and (c) are equivalent. Clearly (d) implies (c), and (a) and (d) are equivalent by the classification theorem for AF-algebras (Theorem 7.3.4). It remains to show that (c) implies (d).

If B is a UHF-algebra and if g is in $K_0(B)$, then g belongs to $K_0(B)^+$ if and only if there are numbers k in $\mathbb{Z}^+$ and l in $\mathbb{N}$ such that $lg = k[1_B]_0$. Indeed, $K_0(B)$ is isomorphic to a subgroup of $\mathbb{Q}$ by Lemma 7.4.4, and so $lg = k[1_B]_0$ for some natural number l and some integer k. Since $K_0(B)$ is unperforated (Definition 7.2.5) we find that g is positive if $k \geqslant 0$, and that $-g$ is positive if $k \leqslant 0$. This proves the claim.

It follows from the claim above that if $\alpha \colon K_0(A) \to K_0(A')$ is an isomorphism satisfying $\alpha([1_A]_0) = [1_{A'}]_0$, then automatically $\alpha(K_0(A)^+) = K_0(A')^+$. This proves that (c) implies (d).

(ii). By Proposition 7.4.3 (i), $G = Q(n)$ for some supernatural number $n = \{n_j\}_{j=1}^\infty$. Put

$$k_j = \prod_{i=1}^{j} p_i^{\min\{j, n_i\}}.$$

Then $k_j | k_{j+1}$ for every j, and n is the supernatural number associated to the sequence $\{k_j\}$ as in Lemma 7.4.4. For each j there is a unital *-homomorphism $\varphi_j \colon M_{k_j}(\mathbb{C}) \to M_{k_{j+1}}(\mathbb{C})$. Let A be the inductive limit of the sequence

$$M_{k_1}(\mathbb{C}) \xrightarrow{\varphi_1} M_{k_2}(\mathbb{C}) \xrightarrow{\varphi_2} M_{k_3}(\mathbb{C}) \xrightarrow{\varphi_3} \cdots .$$

Then $(K_0(A), [1]_0)$ is isomorphic to $(Q(n), 1) = (G, 1)$ by Lemma 7.4.4 (ii).

(iii). By (ii) there is a UHF-algebra A with $(K_0(A), [1]_0)$ isomorphic to $(Q(n), 1)$, and by definition, the supernatural number associated to A is n. $\qquad\square$

It follows from Theorem 7.4.5 that there is, up to isomorphism, only one UHF-algebra associated to each supernatural number n. It is common to write M_n or $M_n(\mathbb{C})$ for this UHF-algebra. In particular, M_{2^∞} is the UHF-algebra which is the inductive limit of the sequence

$$\mathbb{C} \to M_2(\mathbb{C}) \to M_4(\mathbb{C}) \to M_8(\mathbb{C}) \to \cdots$$

with the connecting maps given by $a \mapsto \text{diag}(a, a)$.

Example 7.4.6. We show here that there is an uncountable family of pairwise non-isomorphic UHF-algebras. For each non-empty subset I of $\mathbb{N}$ let n_I be the supernatural number $\{n_i\}_{i=1}^{\infty}$, where $n_i = \infty$ if i is in I, and $n_i = 0$ if i is not in I. In other words, using the prime factorization picture,

$$n_I = \prod_{i \in I} p_i^{\infty}.$$

The family $\{n_I\}$ of all supernatural numbers arising in this way is uncountable. Moreover, $kn_I \neq ln_J$ for every pair of natural numbers k, l when $I \neq J$. It follows that the uncountable family of associated UHF-algebras $\{M_{n_I}(\mathbb{C})\}$ are pairwise non-isomorphic. A more elaborate argument shows that $M_k(M_{n_I}(\mathbb{C}))$ and $M_l(M_{n_J}(\mathbb{C}))$ are non-isomorphic for all natural numbers k, l when $I \neq J$.

7.5   Exercises

Exercise 7.1. Write down explicitly the inductive sequence of finite dimensional C^*-algebras represented by the two last mentioned Bratteli diagrams in Example 7.2.3.

Exercise 7.2. Prove the "only if" part of the Effros–Handelman–Shen theorem (Theorem 7.2.6). [Hint: Use Exercise 6.10.]

Exercise 7.3. Let A be a C^*-algebra and let $\{A_n\}_{n=1}^{\infty}$ be an increasing sequence of sub-C^*-algebras of A whose union is dense in A. Show that

$$I = \overline{\bigcup_{n=1}^{\infty} A_n \cap I}$$

for every closed two-sided ideal I in A. [Hint: Denote the set on the right-hand side by J. Show that J is a closed two-sided ideal in A which is contained in I. Conclude that there is a *-homomorphism $\varphi \colon A/J \to A/I$ making the

diagram

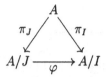

commutative, where π_J and π_I are the quotient mappings. Show that the restriction of φ to $\pi_J(A_n)$ is injective and hence isometric for each n, and conclude from this that φ is injective.]

Show that the AF-algebra represented by the first Bratteli diagram in Example 7.2.3 is simple, and show that the AF-algebra represented by the second Bratteli diagram (the Pascal triangle) in that example is not simple. Show that every UHF-algebra is simple.

Exercise 7.4. Show that $A \oplus B$ has the cancellation property if A and B both have the cancellation property. Show that the inductive limit of a sequence $A_1 \to A_2 \to A_3 \to \cdots$ has the cancellation property if each A_n has the cancellation property. Conclude that every AF-algebra has the cancellation property.

Exercise 7.5. Let A be an AF-algebra. Show that A is stably finite, see Definition 5.1.1. [Hint: First observe that $M_n(\widetilde{A})$ is an AF-algebra whenever A is an AF-algebra and n is a natural number. Conclude that it suffices to show that every unital AF-algebra A is finite. Take an isometry s in A and find x in some finite dimensional sub-C^*-algebra of A with $\|s - x\| < 1$. Show that x is right-invertible, and conclude that s is right-invertible, and hence invertible.]

Exercise 7.6. Let A and B be unital AF-algebras, and let $\varphi, \psi \colon A \to B$ be unit preserving *-homomorphisms with $K_0(\varphi) = K_0(\psi)$. Show that there is a sequence $\{u_n\}_{n=1}^{\infty}$ of unitaries in B such that $(\mathrm{Ad}\, u_n \circ \varphi)(a) \to \psi(a)$ for every a in A. [Hint: Show first, using Exercise 6.7, that there is an increasing sequence of finite dimensional sub-C^*-algebras $\{A_n\}$ of A whose union is dense in A, and such that each A_n contains the unit of A. Show next that there is a unitary u_n in B such that $(\mathrm{Ad}\, u_n \circ \varphi)(a) = \psi(a)$ for every a in A_n, using Lemma 7.3.2 and Exercise 7.4.]

Exercise 7.7. Let A and B be unital AF-algebras, and let $\alpha \colon K_0(A) \to K_0(B)$ be a positive group homomorphism such that $\alpha([1_A]_0) = [1_B]_0$. Show that there is a unital *-homomorphism $\varphi \colon A \to B$ with $K_0(\varphi) = \alpha$. [Hint:

By Exercise 7.4, A has the cancellation property. Find a sequence

$$A_1 \xrightarrow{\lambda_1} A_2 \xrightarrow{\lambda_2} A_3 \longrightarrow \cdots$$

of finite dimensional C^*-algebras with unital connecting *-homomorphisms λ_n, and with inductive limit $(A, \{\mu_n\})$. Then use Lemma 7.3.2 (i) and (ii) to find *-homomorphisms $\varphi_n \colon A_n \to B$ such that $K_0(\varphi_n) = \alpha \circ K_0(\mu_n)$ and $\varphi_{n+1} \circ \lambda_n = \varphi_n$ for all n.]

Exercise 7.8. Let A be a unital AF-algebra. Let here $\text{Aut}(K_0(A))$ denote the set of automorphisms α on $K_0(A)$ satisfying $\alpha(K_0(A)^+) = K_0(A)^+$ and $\alpha([1_A]_0) = [1_A]_0$. Show that the sequence

$$\{1\} \longrightarrow \overline{\text{Inn}}(A) \xrightarrow{\iota} \text{Aut}(A) \xrightarrow{K_0} \text{Aut}(K_0(A)) \longrightarrow \{1\}$$

is exact, where $\{1\}$ denotes the trivial group with one element, where $\overline{\text{Inn}}(A)$ is the approximately inner automorphisms on A (defined in Exercise 4.10), and where ι is the inclusion mapping. [Hint: Use Exercise 4.10, Exercise 7.6, and the classification theorem for AF-algebras (Theorem 7.3.4).]

Exercise 7.9. Let $G = \mathbb{Q} \oplus \mathbb{Q}$ and set

$$G^+ = \{(x, y) \in \mathbb{Q} \oplus \mathbb{Q} : x > 0, \ y > 0\} \cup \{(0, 0)\},$$

see Example 7.2.7 (iv). Show that (G, G^+) is a dimension group and that every non-zero element in G^+ is an order unit for G.

Justify that there is a unital AF-algebra A with $(K_0(A), K_0(A)^+, [1_A]_0)$ isomorphic to $(G, G^+, (1,1))$. In the notation of Exercise 7.8 show that the group $\text{Aut}(K_0(A))$ has two elements. Finally, show that not all automorphisms on A are approximately inner.

Exercise 7.10. Show that one cannot find pairwise orthogonal projections p_1, p_2, p_3 in the UHF-algebra M_{2^∞} such that $p_1 \sim p_2 \sim p_3$ and $p_1 + p_2 + p_3 = 1$.

Exercise 7.11. Let $n = \{n_j\}$ and $n' = \{n'_j\}$ be supernatural numbers. Write $n|n'$ if $n_j \leqslant n'_j$ for every j.

(i) Show that $Q(n) \subseteq Q(n')$ if and only if $n|n'$.

(ii) Show that there is a unital *-homomorphism $\varphi \colon M_n \to M_{n'}$ if and only if $n|n'$. [Hint: Use Exercise 7.7.]

Chapter 8

The Functor K_1

The group $K_1(A)$ of a C^*-algebra A is defined to be the set of homotopy
equivalence classes of unitary elements in matrix algebras over the unitization
of A. It is shown that K_1 is a half exact, homotopy invariant functor. It will
be shown in Chapter 10 that $K_1(A)$ is naturally isomorphic to $K_0(SA)$, and
one can then use this to conclude that many results that hold for K_0 also hold
for K_1. We shall on this account state some results about K_1 whose proofs
are postponed to (exercises in) Chapter 10.

The K_1-group of an Abelian C^*-algebra is partially described through
determinants, and it is in particular shown that $K_1(A)$ is non-zero if A is a
unital Abelian C^*-algebra for which $\mathcal{U}(A)$ is not connected. This fact is used
to show that $K_1(C(\mathbb{T}))$ is non-zero.

8.1   Definition of the K_1-group

Definition 8.1.1. Let A be a unital C^*-algebra, and let as usual $\mathcal{U}(A)$ denote
its group of unitary elements. Set

$$\mathcal{U}_n(A) = \mathcal{U}(M_n(A)), \qquad \mathcal{U}_\infty(A) = \bigcup_{n=1}^{\infty} \mathcal{U}_n(A).$$

Define a binary operation $\oplus$ on $\mathcal{U}_\infty(A)$ by

$$u \oplus v = \begin{pmatrix} u & 0 \\ 0 & v \end{pmatrix} \in \mathcal{U}_{n+m}(A), \qquad u \in \mathcal{U}_n(A),\ v \in \mathcal{U}_m(A).$$

Define a relation $\sim_1$ on $\mathcal{U}_\infty(A)$ as follows. For u in $\mathcal{U}_n(A)$ and v in $\mathcal{U}_m(A)$,
write $u \sim_1 v$ if there exists a natural number $k \geqslant \max\{m,n\}$ such that

$u \oplus 1_{k-n} \sim_h v \oplus 1_{k-m}$ in $\mathcal{U}_k(A)$, where 1_r is the unit in $M_r(A)$ (and with the convention that $w \oplus 1_0 = w$ for all w in $\mathcal{U}_\infty(A)$).

Lemma 8.1.2. *Let A be a unital C^*-algebra. Then*

(i) $\sim_1$ *is an equivalence relation on $\mathcal{U}_\infty(A)$,*

(ii) $u \sim_1 u \oplus 1_n$ *for all u in $\mathcal{U}_\infty(A)$ and n in $\mathbb{N}$,*

(iii) $u \oplus v \sim_1 v \oplus u$ *for all u, v in $\mathcal{U}_\infty(A)$,*

(iv) *if u, u', v, v' belong to $\mathcal{U}_\infty(A)$, $u \sim_1 u'$, and $v \sim_1 v'$, then $u \oplus v \sim_1 u' \oplus v'$,*

(v) *if u, v both belong to $\mathcal{U}_n(A)$ for some n, then $uv \sim_1 vu \sim_1 u \oplus v$,*

(vi) $(u \oplus v) \oplus w = u \oplus (v \oplus w)$ *for all u, v, w in $\mathcal{U}_\infty(A)$.*

Proof. Parts (i), (ii), and (vi) are trivial, and (v) follows from the Whitehead lemma (Lemma 2.1.5).

For the proof of (iii), let u in $\mathcal{U}_n(A)$ and v in $\mathcal{U}_m(A)$ be given, and put

$$z = \begin{pmatrix} 0 & 1_m \\ 1_n & 0 \end{pmatrix} \in \mathcal{U}_{n+m}(A).$$

Then $v \oplus u = z(u \oplus v)z^* \sim_1 z^*z(u \oplus v) = u \oplus v$ by (v).

To prove (iv) it is sufficient to show that

(α) $(u \oplus 1_k) \oplus (v \oplus 1_l) \sim_1 u \oplus v$ for all u, v in $\mathcal{U}_\infty(A)$ and all k, l in $\mathbb{N}$,

(β) $u \sim_h u'$ and $v \sim_h v'$ imply that $u \oplus v \sim_h u' \oplus v'$ for all u, u' in $\mathcal{U}_n(A)$ and all v, v' in $\mathcal{U}_m(A)$.

Statement (α) follows from (ii), (iii), and (vi). To see (β), let $t \mapsto u_t$ and $t \mapsto v_t$ be continuous paths of unitaries with $u_0 = u$, $u_1 = u'$, $v_0 = v$, and $v_1 = v'$. Then $t \mapsto u_t \oplus v_t$ is a continuous path of unitaries from $u \oplus v$ to $u' \oplus v'$. $\square$

Definition 8.1.3 (The K_1-group). For each C^*-algebra A define

$$K_1(A) = \mathcal{U}_\infty(\tilde{A})/\sim_1.$$

Let $[u]_1$ in $K_1(A)$ denote the equivalence class containing u in $\mathcal{U}_\infty(\tilde{A})$. Define a binary operation $+$ on $K_1(A)$ by $[u]_1 + [v]_1 = [u \oplus v]_1$, where u, v belong to $\mathcal{U}_\infty(\tilde{A})$.

Lemma 8.1.2 shows that $+$ is well-defined, commutative, associative, has zero element $[1]_1$ ($= [1_n]_1$ for each n in $\mathbb{N}$), and that

$$0 = [1_n]_1 = [uu^*]_1 = [u]_1 + [u^*]_1$$

for each u in $\mathcal{U}_n(\widetilde{A})$. This shows that $(K_1(A), +)$ is an Abelian group, and $-[u]_1 = [u^*]_1$ for all u in $\mathcal{U}_\infty(\widetilde{A})$.

We record the following standard picture and universal property of K_1. The standard picture is really just a restatement of the definition of K_1.

Proposition 8.1.4 (The standard picture of K_1). *Let A be a C^*-algebra. Then*

(8.1) $$K_1(A) = \{[u]_1 : u \in \mathcal{U}_\infty(\widetilde{A})\},$$

and the map $[\,\cdot\,]_1 : \mathcal{U}_\infty(\widetilde{A}) \to K_1(A)$ has the following properties:

(i) $[u \oplus v]_1 = [u]_1 + [v]_1,$

(ii) $[1]_1 = 0,$

(iii) *if u, v belong to $\mathcal{U}_n(\widetilde{A})$ and $u \sim_h v$, then $[u]_1 = [v]_1,$*

(iv) *if u, v belong to $\mathcal{U}_n(\widetilde{A})$, then $[uv]_1 = [vu]_1 = [u]_1 + [v]_1,$*

(v) *for u, v in $\mathcal{U}_\infty(\widetilde{A})$, $[u]_1 = [v]_1$ if and only if $u \sim_1 v$.*

Proof. The properties (i), (ii), (v) and equation (8.1) are immediate consequences of Definition 8.1.3, (iii) follows from (v), and Lemma 8.1.2 (v) shows that (iv) holds. □

Proposition 8.1.5 (Universal property of K_1). *Let A be a C^*-algebra, let G be an Abelian group, and let $\nu : \mathcal{U}_\infty(\widetilde{A}) \to G$ be a map with the following properties:*

(i) $\nu(u \oplus v) = \nu(u) + \nu(v),$

(ii) $\nu(1) = 0,$

(iii) *if u, v belong to $\mathcal{U}_n(\widetilde{A})$ and $u \sim_h v$, then $\nu(u) = \nu(v)$.*

Then there exists a unique group homomorphism $\alpha\colon K_1(A) \to G$ *making the diagram*

(8.2)

$$
\begin{array}{ccc}
 & \mathcal{U}_\infty(\widetilde{A}) & \\
{\scriptstyle [\,\cdot\,]_1}\Big\downarrow & \searrow{\scriptstyle \nu} & \\
K_1(A) & \xrightarrow[\alpha]{} & G
\end{array}
$$

commutative.

Proof. We show first that if u in $\mathcal{U}_n(\widetilde{A})$ and v in $\mathcal{U}_m(\widetilde{A})$ are such that $u \sim_1 v$, then $\nu(u) = \nu(v)$. Find to this end an integer $k \geqslant \max\{n, m\}$ with $u \oplus 1_{k-n} \sim_h v \oplus 1_{k-m}$ in $\mathcal{U}_k(\widetilde{A})$. Use (i) and (ii) to conclude that $\nu(1_r) = 0$ for all r in $\mathbb{N}$. Then (i) and (iii) imply that

$$\nu(u) = \nu(u \oplus 1_{k-n}) = \nu(v \oplus 1_{k-m}) = \nu(v).$$

Consequently, there is a map $\alpha\colon K_1(A) \to G$ making the diagram (8.2) commutative. The computation

$$
\begin{aligned}
\alpha([u]_1 + [v]_1) &= \alpha([u \oplus v]_1) = \nu(u \oplus v) = \nu(u) + \nu(v) \\
&= \alpha([u]_1) + \alpha([v]_1)
\end{aligned}
$$

shows that α is a group homomorphism. The uniqueness of α follows from the surjectivity of $[\,\cdot\,]_1$.   □

It would be natural to define $K_1(A)$ of a *unital C^*-algebra* to be $\mathcal{U}_\infty(A)/\sim_1$, and, indeed, as the proposition below shows, this is consistent with Definition 8.1.3.

Proposition 8.1.6. *Let A be a unital C^*-algebra. Then there is an isomorphism* $\rho\colon K_1(A) \to \mathcal{U}_\infty(A)/\sim_1$ *making the diagram*

(8.3)

$$
\begin{array}{ccc}
\mathcal{U}_\infty(\widetilde{A}) & \xrightarrow{\ \mu\ } & \mathcal{U}_\infty(A) \\
{\scriptstyle [\,\cdot\,]_1}\Big\downarrow & & \Big\downarrow \\
K_1(A) & \xrightarrow[\rho]{} & \mathcal{U}_\infty(A)/\sim_1
\end{array}
$$

commutative.

The map μ in the diagram (8.3) is defined as follows. As in Paragraph 1.1.6, $\widetilde{A} = A + \mathbb{C}f$, where $f = 1_{\widetilde{A}} - 1_A$. The map $\mu \colon \widetilde{A} \to A$ given by $\mu(a + \alpha f) = a$, for a in A and α in $\mathbb{C}$, is a unital *-homomorphism. Extend μ to a unital *-homomorphism $M_n(\widetilde{A}) \to M_n(A)$ for each n in $\mathbb{N}$ (see Section 1.3), and obtain in this way a map $\mu \colon \mathcal{U}_\infty(\widetilde{A}) \to \mathcal{U}_\infty(A)$. See also Exercise 8.5.

Proof of proposition. First note that $\mu \colon \mathcal{U}_\infty(\widetilde{A}) \to \mathcal{U}_\infty(A)$ is surjective. Hence it suffices to show that

(α) $\mu(u) \sim_1 \mu(v)$ if and only if $u \sim_1 v$ for all u, v in $\mathcal{U}_\infty(\widetilde{A})$,

(β) $\mu(u \oplus v) = \mu(u) \oplus \mu(v)$ for all u, v in $\mathcal{U}_\infty(\widetilde{A})$.

That (β) holds follows immediately from the definition of the map μ. To show (α) it suffices to show that

(α') $\mu(u) \sim_h \mu(v)$ if and only if $u \sim_h v$ for all u, v in $\mathcal{U}_n(\widetilde{A})$ and all n in $\mathbb{N}$.

The "if" implication in (α') is trivial. To prove the "only if" implication, let u, v belong to $\mathcal{U}_n(\widetilde{A})$ and suppose that $\mu(u) \sim_h \mu(v)$ in $\mathcal{U}_n(A)$. By the definition of μ we can find u_0, v_0 in $\mathcal{U}_n(\mathbb{C}f)$ such that $u = \mu(u) + u_0$ and $v = \mu(v) + v_0$. Lemma 2.1.3 (ii) shows that $u_0 \sim_h v_0$ in $\mathcal{U}_n(\mathbb{C}f)$. This proves that $u \sim_h v$ in $\mathcal{U}_n(\widetilde{A})$. Indeed, take continuous paths $t \mapsto a_t$ and $t \mapsto b_t$ of unitaries in $M_n(A)$, respectively, in $M_n(\mathbb{C}f)$, with $a_0 = \mu(u)$, $a_1 = \mu(v)$, $b_0 = u_0$, and $b_1 = v_0$. Then $t \mapsto a_t + b_t$ is a continuous path in $\mathcal{U}_n(\widetilde{A})$ with $a_0 + b_0 = u$ and $a_1 + b_1 = v$. $\qquad\square$

When A is unital we shall often identify $K_1(A)$ with $\mathcal{U}_\infty(A)/\sim_1$ via the isomorphism ρ from Proposition 8.1.6. If u is a unitary element in $\mathcal{U}_\infty(A)$, then $[u]_1$ will denote the element in $K_1(A)$ it represents under this identification.

It is an immediate consequence of Proposition 8.1.6 that

(8.4) $$K_1(A) \cong K_1(\widetilde{A})$$

for every C^*-algebra A.

Remark 8.1.7. Proposition 2.1.8 shows that in a natural way we can extend the map $[\,\cdot\,]_1 \colon \mathcal{U}_\infty(\widetilde{A}) \to K_1(A)$ to a map

$$[\,\cdot\,]_1 \colon \mathrm{GL}_\infty(\widetilde{A}) \to K_1(A),$$

where $\mathrm{GL}_n(\widetilde{A}) = \mathrm{GL}(M_n(\widetilde{A}))$ and $\mathrm{GL}_\infty(\widetilde{A}) = \bigcup_{n=1}^\infty \mathrm{GL}_n(\widetilde{A})$. Moreover, if A is unital, then we can extend the map $[\,\cdot\,]_1 \colon \mathcal{U}_\infty(A) \to K_1(A)$ to a map

$$[\,\cdot\,]_1 \colon \mathrm{GL}_\infty(A) \to K_1(A).$$

These extensions are defined as follows. For a in $\mathrm{GL}_n(\widetilde{A})$ (respectively a in $\mathrm{GL}_n(A)$), let $[a]_1 = [u]_1$, where u is any unitary in $\mathcal{U}_n(\widetilde{A})$ (respectively in $\mathcal{U}_n(A)$) with the property that $a \sim_h u$ in $\mathrm{GL}_n(\widetilde{A})$ (respectively in $\mathrm{GL}_n(A)$). By Proposition 2.1.8 one can choose u to be $a(a^*a)^{-1/2}$, the unitary in the polar decomposition for a.

Example 8.1.8. $K_1(\mathbb{C}) = K_1(M_n(\mathbb{C})) = 0$. More generally, $K_1(B(H)) = 0$ for each Hilbert space H.

Proof. It is shown in Corollary 2.1.4 that the unitary group of $M_k(M_n(\mathbb{C})) = M_{kn}(\mathbb{C})$ is connected for every n and every k. This implies that $\mathcal{U}_n(M_n(\mathbb{C}))/\sim_1$ is the trivial group with one element. From the description of K_1 of a unital C^*-algebra in Proposition 8.1.6 we conclude that $K_1(M_n(\mathbb{C})) = 0$.

Let now H be a Hilbert space. We show that $u \sim_h 1_n$ for every unitary u in $\mathcal{U}_n(B(H))$, and this will imply that $u \sim_1 1$, and hence that $\mathcal{U}_\infty(B(H))/\sim_1$ is zero. As above, this will show that $K_1(B(H)) = 0$.

Define $\varphi \colon \mathbb{T} \to [0, 2\pi)$ by

$$\varphi(e^{i\theta}) = \theta, \quad 0 \leqslant \theta < 2\pi.$$

Then φ is a bounded Borel measurable map, and $z = \exp(i\varphi(z))$ for each z in $\mathbb{T}$. Let u be a unitary in $\mathcal{U}_n(B(H)) = \mathcal{U}(B(H^n))$. Then $\varphi(u) = \varphi(u)^*$ in $B(H^n)$ and $u = \exp(i\varphi(u))$. Apply Proposition 2.1.6 to conclude that $u \sim_h 1$.   □

8.2   Functoriality of K_1

8.2.1 Functoriality of K_1. Let A and B be C^*-algebras, and let $\varphi \colon A \to B$ be a *-homomorphism. Then φ induces a unital *-homomorphism $\widetilde{\varphi} \colon \widetilde{A} \to \widetilde{B}$ that extends to a unital *-homomorphism $\widetilde{\varphi} \colon M_n(\widetilde{A}) \to M_n(\widetilde{B})$ for each n in $\mathbb{N}$, see Section 1.3. This gives a map $\widetilde{\varphi} \colon \mathcal{U}_\infty(\widetilde{A}) \to \mathcal{U}_\infty(\widetilde{B})$, and we define $\nu \colon \mathcal{U}_\infty(\widetilde{A}) \to K_1(B)$ by $\nu(u) = [\widetilde{\varphi}(u)]_1$ for each u in $\mathcal{U}_\infty(\widetilde{A})$. It is straightforward to verify that ν fulfills (i), (ii), and (iii) in Proposition 8.1.5, and hence there is precisely one group homomorphism $K_1(\varphi) \colon K_1(A) \to K_1(B)$

with the property

(8.5) $$K_1(\varphi)([u]_1) = [\widetilde{\varphi}(u)]_1, \qquad u \in \mathcal{U}_\infty(\widetilde{A}).$$

If A and B are unital C^*-algebras and $\varphi\colon A \to B$ is a unital *-homomorphism, then $K_1(\varphi)([u]_1) = [\varphi(u)]_1$ for all u in $\mathcal{U}_\infty(A)$. (See Exercise 8.5.)

The proposition below shows that K_1 is a homotopy invariant functor which preserves zero objects. (See Paragraph 3.2.1.)

Proposition 8.2.2 (Functoriality of K_1). *Let A, B, and C be C^*-algebras. Then*

(i) $K_1(\mathrm{id}_A) = \mathrm{id}_{K_1(A)}$,

(ii) $K_1(\psi \circ \varphi) = K_1(\psi) \circ K_1(\varphi)$, *if* $\varphi\colon A \to B$ *and* $\psi\colon B \to C$ *are* *-homomorphisms.*

In particular K_1 is a functor. Furthermore,

(iii) $K_1(\{0\}) = \{0\}$,

(iv) $K_1(0_{B,A}) = 0_{K_1(B),K_1(A)}$,

(v) *if $\varphi, \psi\colon A \to B$ are homotopic *-homomorphisms, then $K_1(\varphi) = K_1(\psi)$,*

(vi) *if A and B are homotopy equivalent, then $K_1(A)$ is isomorphic to $K_1(B)$. More specifically, if*

$$A \xrightarrow{\varphi} B \xrightarrow{\psi} A$$

is a homotopy, then

$$K_1(\varphi)\colon K_1(A) \to K_1(B) \quad and \quad K_1(\psi)\colon K_1(B) \to K_1(A)$$

are isomorphisms, and $K_1(\varphi)^{-1} = K_1(\psi)$.

Proof. Use (8.5) and the facts that $(\mathrm{id}_A)^{\sim} = \mathrm{id}_{\widetilde{A}}$ and $(\psi \circ \varphi)^{\sim} = \widetilde{\psi} \circ \widetilde{\varphi}$ to see (i) and (ii).

(iii). It follows from Proposition 8.1.6 that $K_1(A)$ is isomorphic to $K_1(\widetilde{A})$ for every C^*-algebra A. In particular we get that $K_1(\{0\})$ is isomorphic to $K_1(\mathbb{C})$. It is shown in Example 8.1.8 that $K_1(\mathbb{C}) = 0$.

(iv). The zero homomorphism $0_{B,A}$ is the composition of the maps $A \to$ $\{0\} \to B$. Hence (iv) follows from (iii) and (ii).

(v). Let $\varphi_t \colon A \to B$, $t \in [0,1]$, be a path of *-homomorphisms such that $\varphi_0 = \varphi$, $\varphi_1 = \psi$, and $t \mapsto \varphi_t(a)$ is continuous for all a in A. The induced *-homomorphisms $\widetilde{\varphi}_t \colon M_n(\widetilde{A}) \to M_n(\widetilde{B})$ are unital, and the maps $t \to \widetilde{\varphi}_t(a)$ are continuous for every a in $M_n(\widetilde{A})$. Hence $\widetilde{\varphi}(u) = \widetilde{\varphi}_0(u) \sim_h \widetilde{\varphi}_1(u) = \widetilde{\psi}(u)$ in $\mathcal{U}_n(\widetilde{B})$ for all u in $\mathcal{U}_n(\widetilde{A})$. It follows that

$$K_1(\varphi)([u]_1) = [\widetilde{\varphi}(u)]_1 = [\widetilde{\psi}(u)]_1 = K_1(\psi)([u]_1),$$

which proves (v).

Part (vi) is a consequence of (i), (ii), and (v). $\square$

Lemma 8.2.3. *Let A and B be C^*-algebras, let $\varphi \colon A \to B$ be a *-homomorphism, and let g be an element in the kernel of $K_1(\varphi)$. Then*

(i) *there is a unitary u in $\mathcal{U}_\infty(\widetilde{A})$ such that $g = [u]_1$ and $\widetilde{\varphi}(u) \sim_h 1$,*

(ii) *if φ is surjective, then there is a unitary u in $\mathcal{U}_\infty(\widetilde{A})$ such that $g = [u]_1$ and $\widetilde{\varphi}(u) = 1$.*

Proof. (i). Choose a unitary v in $\mathcal{U}_m(\widetilde{A})$ with $g = [v]_1$. Then $[\widetilde{\varphi}(v)]_1 = 0 = [1_m]_1$, and hence there is an integer $n \geqslant m$ such that

$$\widetilde{\varphi}(v) \oplus 1_{n-m} \sim_h 1_m \oplus 1_{n-m} = 1_n.$$

Put $u = v \oplus 1_{n-m}$. Then $[u]_1 = [v]_1 = g$ and $\widetilde{\varphi}(u) = \widetilde{\varphi}(v) \oplus 1_{n-m} \sim_h 1_n$.

(ii). Use (i) to find v in $\mathcal{U}_n(\widetilde{A})$ with $g = [v]_1$ and $\widetilde{\varphi}(v) \sim_h 1$. Lemma 2.1.7 gives the existence of a unitary w in $\mathcal{U}_n(\widetilde{A})$ fulfilling $\widetilde{\varphi}(w) = \widetilde{\varphi}(v)$ and $w \sim_h 1$. Then $u = w^*v$ has the desired properties. $\square$

Proposition 8.2.4 (Half exactness of K_1). *Let*

$$0 \longrightarrow I \overset{\varphi}{\longrightarrow} A \overset{\psi}{\longrightarrow} B \longrightarrow 0$$

be a short exact sequence of C^*-algebras. Then the sequence

$$K_1(I) \xrightarrow{\;K_1(\varphi)\;} K_1(A) \xrightarrow{\;K_1(\psi)\;} K_1(B),$$

is exact, that is, $\mathrm{Im}(K_1(\varphi)) = \mathrm{Ker}(K_1(\psi))$.

Proof. Because K_1 is a functor that maps zero morphisms to zero morphisms (Proposition 8.2.2) and $\psi \circ \varphi = 0$, we conclude that $K_1(\psi) \circ K_1(\varphi) = 0$. This shows that $\text{Im}(K_1(\varphi))$ is contained in $\text{Ker}(K_1(\psi))$.

To prove the reverse inclusion, let g be in the kernel of $K_1(\psi)$. Use Lemma 8.2.3 (ii) to find a unitary u in $\mathcal{U}_n(\widetilde{A})$ with $g = [u]_1$ and $\widetilde{\psi}(u) = 1$. Next apply Lemma 4.3.1 (ii) to get a unitary v in $\mathcal{U}_n(\widetilde{I})$ such that $\widetilde{\varphi}(v) = u$. Then $[v]_1$ belongs to $K_1(I)$, and $K_1(\varphi)([v]_1) = [\widetilde{\varphi}(v)]_1 = [u]_1 = g$. $\qquad\square$

The functor K_1 is split exact and preserves direct sums of C^*-algebras. These facts can be proved in the same way as for the functor K_0 in Propositions 4.3.3 and 4.3.4 (see Exercise 8.16). They also *follow* from these results and the isomorphism $K_1(A) \cong K_0(SA)$ which will be established in Chapter 10 (see Exercise 10.5). We shall for this reason omit the proofs for now.

Proposition 8.2.5 (Split exactness of K_1). *Let*

$$0 \longrightarrow I \xrightarrow{\varphi} A \underset{\lambda}{\overset{\psi}{\rightleftarrows}} B \longrightarrow 0$$

be a split exact sequence of C^*-algebras. Then

$$0 \longrightarrow K_1(I) \xrightarrow{K_1(\varphi)} K_1(A) \underset{K_1(\lambda)}{\overset{K_1(\psi)}{\rightleftarrows}} K_1(B) \longrightarrow 0$$

is a split exact sequence of Abelian groups.

Proposition 8.2.6 (Direct sums). *Let A and B be C^*-algebras. Then*

$$K_1(A \oplus B) \cong K_1(A) \oplus K_1(B).$$

More specifically, if $\iota_A \colon A \to A \oplus B$ and $\iota_B \colon B \to A \oplus B$ are the inclusion mappings, then the map

$$K_1(\iota_A) \oplus K_1(\iota_B) \colon K_1(A) \oplus K_1(B) \to K_1(A \oplus B),$$
$$(g, h) \mapsto K_1(\iota_A)(g) + K_1(\iota_B)(h),$$

is a group isomorphism.

For reference we record the following two results that will be proved in Exercises 10.3 and 10.4 in Chapter 10 (using stability of K_0 and continuity of K_0, respectively, and the isomorphism $K_1(A) \cong K_0(SA)$).

Proposition 8.2.7 (Continuity of K_1). *Let*

$$A_1 \xrightarrow{\varphi_1} A_2 \xrightarrow{\varphi_2} A_3 \longrightarrow \cdots$$

be a sequence of C^*-algebras with inductive limit $(A, \{\mu_n\})$, and let $(G_1, \{\beta_n\})$ be the inductive limit of the sequence

$$K_1(A_1) \xrightarrow{K_1(\varphi_1)} K_1(A_2) \xrightarrow{K_1(\varphi_2)} K_1(A_3) \longrightarrow \cdots .$$

Then there is a group isomorphism $\gamma \colon G_1 \to K_1(A)$ that satisfies $\gamma \circ \beta_n = K_1(\mu_n)$ for each n, and

(i) $K_1(A) = \displaystyle\bigcup_{n=1}^{\infty} K_1(\mu_n)(K_1(A_n)),$

(ii) $\mathrm{Ker}(K_1(\mu_n)) = \displaystyle\bigcup_{n=m+1}^{\infty} \mathrm{Ker}(K_1(\varphi_{m,n}))$ *for each n in $\mathbf{N}$.*

Proposition 8.2.8 (Stability of K_1). *Let A be a C^*-algebra, and let n be a natural number. Let $\lambda_{n,A} \colon A \to M_n(A)$ be the map defined in Proposition 4.3.8, and let $\kappa \colon A \to \mathcal{K}A$ be the map defined in Section 6.4. Then*

$$K_1(\lambda_{n,A}) \colon K_1(A) \to K_1(M_n(A)) \quad \text{and} \quad K_1(\kappa) \colon K_1(A) \to K_1(\mathcal{K}A)$$

are isomorphisms.

Example 8.2.9. With $\mathcal{K}$ being the algebra of all compact operators on a separable Hilbert space, we have $K_1(\mathcal{K}) = 0$.

Proof. From Proposition 8.2.8, $K_1(\mathcal{K}) \cong K_1(\mathbb{C})$, and $K_1(\mathbb{C}) = 0$ by Example 8.1.8. □

8.2.10* K-theory and partial unitaries. One can simultaneously define the two K-groups $K_0(A)$ and $K_1(A)$ of a (unital) C^*-algebra A using so-called *partial unitary* elements: elements that are partial isometries and normal. In other words, an element u is a partial unitary if and only if $uu^* = u^*u$ and uu^* is a projection. All unitary elements and all projections are partial unitary elements. A straightforward calculation (or Exercise 2.6) shows that $u + (1_A - u^*u)$ is unitary when u is a partial unitary element in a unital C^*-algebra A.

We describe, without going into details, how one can construct an Abelian group, here called $K(A)$, for each unital C^*-algebra A using partial unitary elements.

Let $\mathrm{PU}(A)$ denote the set of all partial unitary elements in A. Set

$$\mathrm{PU}_n(A) = \mathrm{PU}(M_n(A)), \qquad \mathrm{PU}_\infty(A) = \bigcup_{n=1}^{\infty} \mathrm{PU}_n(A).$$

Define an operation $\oplus$ on $\mathrm{PU}_\infty(A)$ by $u \oplus v = \mathrm{diag}(u, v)$, and define a relation $\sim$ on $\mathrm{PU}_\infty(A)$ as follows. If u belongs to $\mathrm{PU}_n(A)$ and v belongs to $\mathrm{PU}_m(A)$, then $u \sim v$ if $u \oplus 0_{k-n} \sim_h v \oplus 0_{k-m}$ in $\mathrm{PU}_k(A)$ for some $k \geqslant \max\{n, m\}$.

There is a well-defined operation $+$ on the quotient $\mathrm{PU}_\infty(A)/\sim$ satisfying

$$\langle u \rangle + \langle v \rangle = \langle u \oplus v \rangle, \qquad u, v \in \mathrm{PU}_\infty(A),$$

where $\langle u \rangle$ is the equivalence class in $\mathrm{PU}_\infty(A)/\sim$ containing u. The operation $+$ is commutative and associative.

Define $K(A)$ to be the Grothendieck group of the Abelian semigroup $(\mathrm{PU}_\infty(A)/\sim, +)$. Let $\gamma\colon \mathrm{PU}_\infty(A)/\sim \to K(A)$ be the Grothendieck map. Let $[u]$ denote the class of u in $K(A)$, i.e., $[u] = \gamma(\langle u \rangle)$. One has a universal property for K which is analogous to the universal property of K_0 and K_1 (Propositions 3.1.8 and 8.1.5): Let G be an Abelian group, and let $\nu\colon \mathrm{PU}_\infty(A) \to G$ be a map that satisfies the conditions (i) $\nu(u \oplus v) = \nu(u) + \nu(v)$, (ii) $\nu(0) = 0$, and (iii) $\nu(u) = \nu(v)$ if u, v belong to $\mathrm{PU}_n(A)$ and $u \sim_h v$ in $\mathrm{PU}_n(A)$ for some n. Then there is a group homomorphism $\alpha\colon K(A) \to G$ that satisfies $\alpha([u]) = \nu(u)$ for all u in $\mathrm{PU}_\infty(A)$.

Using the universal property of K, one obtains group homomorphisms $\alpha_0\colon K(A) \to K_0(A)$ and $\alpha_1\colon K(A) \to K_1(A)$ satisfying

$$\alpha_0([u]) = [u^*u]_0, \qquad \alpha_1([u]) = [u + (1_n - u^*u)]_1, \quad u \in \mathrm{PU}_n(A).$$

Define $\alpha\colon K(A) \to K_0(A) \oplus K_1(A)$ by $\alpha(g) = (\alpha_0(g), \alpha_1(g))$. Then α is an isomorphism and hence

$$K(A) \cong K_0(A) \oplus K_1(A).$$

To see that α is surjective, take a natural number n, a projection p in $\mathcal{P}_n(A)$, and a unitary u in $\mathcal{U}_n(A)$. Then p and u belong to $\mathrm{PU}_n(A)$ and

$$\alpha([p]) = ([p]_0, [p + (1_n - p)]_1) = ([p]_0, 0),$$
$$\alpha([u]) = ([u^*u]_0, [u]_1) = ([1_n]_0, [u]_1).$$

Since elements of the forms $([p]_0, 0)$ and $([1_n]_0, [u]_1)$ generate the Abelian group $K_0(A) \oplus K_1(A)$, we see that α is surjective.

To see that α is injective, take an element g in the kernel of α. Find u in $\mathrm{PU}_k(A)$ and v in $\mathrm{PU}_l(A)$ such that $g = [u] - [v]$. Upon choosing an integer $n \geqslant \max\{k, l\}$ and replacing u and v by $u \oplus 0_{n-k}$ and $v \oplus 0_{n-l}$, respectively, we can assume that u and v both belong to $\mathrm{PU}_n(A)$. Now, $0 = \alpha_0(g) = [u^*u]_0 - [v^*v]_0$. Hence u^*u is stably equivalent to v^*v, and we can therefore find a natural number m such that $u^*u \oplus 1_m$ and $v^*v \oplus 1_m$ are Murray–von Neumann equivalent. Use Proposition 2.2.8 (i) to find a unitary z in $M_{2(n+m)}(A)$ with

$$z^*(u^*u \oplus 1_m \oplus 0_{n+m})z = v^*v \oplus 1_m \oplus 0_{n+m}.$$

A version of the Whitehead lemma (see the proof of Proposition 2.2.8 (ii)) shows that $w^*xw \oplus 0_n \sim_h x \oplus 0_n$ in $\mathrm{PU}_{2n}(A)$ for each x in $\mathrm{PU}_n(A)$ and for each w in $\mathcal{U}_n(A)$. Hence, upon replacing u and v by $z^*(u \oplus 1_m \oplus 0_{2m})z$ and $v \oplus 1_m \oplus 0_{2m}$, respectively, and n by $2(n + m)$, we obtain that $g = [u] - [v]$ for some u, v in $\mathrm{PU}_n(A)$ with $u^*u = v^*v$. Finally, by replacing u and v by $u + (1_n - u^*u)$ and $v + (1_n - v^*v)$, respectively, we can assume that $g = [u] - [v]$, where u, v are unitary elements in $\mathcal{U}_n(A)$. Now, $0 = \alpha_1(g) = [u]_1 - [v]_1$, and so $u \sim_1 v$. This entails that $u \oplus 1_r \sim_h v \oplus 1_r$ in $\mathcal{U}_{n+r}(A)$ for some natural number r. Since $\mathcal{U}_{n+r}(A)$ is contained in $\mathrm{PU}_{n+r}(A)$ we conclude that $u \oplus 1_r \sim_h v \oplus 1_r$ in $\mathrm{PU}_{n+r}(A)$. Hence

$$g = [u] - [v] = [u] + [1_r] - [v] - [1_r] = [u \oplus 1_r] - [v \oplus 1_r] = 0,$$

which shows that α is injective.

8.3* K_1-groups and determinants

For each unital C^*-algebra A there is a group homomorphism ω making the diagram

$$
\begin{array}{ccc}
\mathcal{U}(A) & & \\
\big\downarrow & \searrow^{[\cdot]_1} & \\
\mathcal{U}(A)/\mathcal{U}_0(A) & \xrightarrow[\omega]{} & K_1(A)
\end{array}
$$

commutative. That ω exists follows from the fact that $[u]_1 = 0$ for each u in $\mathcal{U}_0(A)$. The map ω is in general neither injective nor surjective, but in some important cases ω is actually an isomorphism. In this section we shall investigate the map ω for unital Abelian C^*-algebras.

Let A be an Abelian C^*-algebra. For each natural number n define the *determinant* $D: M_n(A) \to A$ by

$$D\begin{pmatrix} a_{11} & \cdots & a_{1n} \\ \vdots & \ddots & \vdots \\ a_{n1} & \cdots & a_{nn} \end{pmatrix} = \sum_{\sigma \in S_n} \text{sign}(\sigma) \prod_{j=1}^{n} a_{j\sigma(j)},$$

where S_n is the group of all permutations of $\{1, 2, \ldots, n\}$. If $A = \mathbb{C}$, then D is the usual determinant.

As with the usual determinant (when $A = \mathbb{C}$) the determinant D has the following properties:

- $D(ab) = D(a)D(b)$ for all a, b in $M_n(A)$,

- $D\begin{pmatrix} a & 0 \\ 0 & b \end{pmatrix} = D(a)D(b)$ for all a in $M_n(A)$ and b in $M_m(A)$,

- $D(a^*) = D(a)^*$ for all a in $M_n(A)$,

- $D(a) = a$ for all a in A,

- $D: M_n(A) \to A$ is continuous for every n.

Thus, if A is unital (and still Abelian), then D maps $\mathcal{U}_\infty(A)$ into $\mathcal{U}(A)$ and

- $D\begin{pmatrix} u & 0 \\ 0 & v \end{pmatrix} = D(u)D(v)$ for u, v in $\mathcal{U}_\infty(A)$,

- if u, v belong to $\mathcal{U}_n(A)$ and $u \sim_h v$, then $D(u) \sim_h D(v)$.

Let $\langle u \rangle$ denote the equivalence class in $\mathcal{U}(A)/\mathcal{U}_0(A)$ of u in $\mathcal{U}(A)$. From the universal property of K_1 (Proposition 8.1.5) applied to the map

$$\nu: \mathcal{U}_\infty(A) \to \mathcal{U}(A)/\mathcal{U}_0(A), \qquad \nu(u) = \langle D(u) \rangle,$$

we get a homomorphism $\Delta: K_1(A) \to \mathcal{U}(A)/\mathcal{U}_0(A)$ with $\Delta([u]_1) = \langle D(u) \rangle$ for each u in $\mathcal{U}_\infty(A)$. In this way we get a commutative diagram:

$$
\begin{array}{ccc}
 & \mathcal{U}(A) & \\
 \swarrow & \downarrow & \searrow \\
\mathcal{U}(A)/\mathcal{U}_0(A) \xrightarrow[\omega]{} & K_1(A) & \xrightarrow[\Delta]{} \mathcal{U}(A)/\mathcal{U}_0(A)
\end{array}
$$

In particular, $\Delta \circ \omega$ is the identity map on $\mathcal{U}(A)/\mathcal{U}_0(A)$. We have thus proved the following proposition.

Proposition 8.3.1. *Let A be a unital Abelian C^*-algebra. Then*

$$(8.6) \qquad 0 \longrightarrow \mathrm{Ker}(\Delta) \overset{\iota}{\longhookrightarrow} K_1(A) \underset{\omega}{\overset{\Delta}{\rightleftarrows}} \mathcal{U}(A)/\mathcal{U}_0(A) \longrightarrow 0$$

is a split exact sequence of Abelian groups.
In particular, $\omega \colon \mathcal{U}(A)/\mathcal{U}_0(A) \to K_1(A)$ is injective, and

$$K_1(A) \cong \mathcal{U}(A)/\mathcal{U}_0(A) \oplus \mathrm{Ker}(\Delta).$$

Note the immediate consequence of Proposition 8.3.1 that $K_1(A)$ is non-zero if A is a unital Abelian C^*-algebra and $\mathcal{U}(A)$ is not connected.

If $A = C(X)$ for some compact Hausdorff space X, then $\mathcal{U}(A)$ is equal to the set of continuous functions $C(X, \mathbb{T})$. The group of homotopy equivalence classes in $C(X, \mathbb{T})$ is called the *cohomotopy group* of X, and it is denoted by $\pi^1(X)$. In other words,

$$\pi^1(X) = \mathcal{U}(C(X))/\mathcal{U}_0(C(X)).$$

The short exact sequence in (8.6) can thus be rewritten as

$$(8.7) \qquad 0 \longrightarrow \mathrm{Ker}(\Delta) \overset{\iota}{\longrightarrow} K_1(C(X)) \underset{\omega}{\overset{\Delta}{\rightleftarrows}} \pi^1(X) \longrightarrow 0.$$

We shall use these facts in the following example.

Example 8.3.2. The cohomotopy group $\pi^1(\mathbb{T})$ is isomorphic to $\mathbb{Z}$. In particular, $K_1(C(\mathbb{T})) \neq 0$.

Proof. For each u in $C(\mathbb{T}, \mathbb{T})$ there is a continuous function $f \colon [0,1] \to \mathbb{R}$ such that

$$(8.8) \qquad u\left(e^{2\pi i t}\right) = e^{2\pi i f(t)}, \quad t \in [0,1].$$

If $f, g \colon [0,1] \to \mathbb{R}$ are continuous and both satisfy (8.8), then $f - g$ is a constant integer. The map $w \colon C(\mathbb{T}, \mathbb{T}) \to \mathbb{Z}$ given by $w(u) = f(1) - f(0)$ is therefore well-defined. The integer $w(u)$ is called the *winding number* of u. The winding number map w is surjective, and it satisfies

- $u \sim_h v$ if and only if $w(u) = w(v)$,

- $w(uv) = w(u) + w(v)$,

whenever u, v belong to $C(\mathbb{T}, \mathbb{T})$. Hence $[u] \mapsto w(u)$ defines an isomorphism $\pi^1(\mathbb{T}) \to \mathbb{Z}$, where $[u]$ denotes the homotopy class in $\pi^1(\mathbb{T})$ of u. □

It will be shown in Example 11.3.4 that $\Delta \colon K_1(C(\mathbb{T})) \to \pi^1(\mathbb{T})$ is an isomorphism, and hence that $K_1(C(\mathbb{T})) \cong \mathbb{Z}$.

8.4 Exercises

Exercise 8.1. Let A be a separable C^*-algebra. Show that $K_1(A)$ is a countable Abelian group.

Exercise 8.2. Show that $K_1(C(X)) = 0$ whenever X is a compact, contractible Hausdorff space.

Exercise 8.3. Let A be an arbitrary C^*-algebra, and let CA denote the cone of A (see Example 4.1.5). Show that $K_1(CA) = 0$.

Exercise 8.4. In this exercise one can assume the fact (to be proved in Chapter 11) that $K_1(C(\mathbb{T}))$ is isomorphic to $\mathbb{Z}$. Determine $K_1(C_0(\mathbb{R}))$, and calculate $K_1(C_0(U))$, where $U = [-3,0] \cup (1,3) \cup (5,6) \cup (8,11] \cup \{15\}$. [Hint: Look at Exercise 4.2.]

Exercise 8.5. Let A and B be unital C^*-algebras, and let $\varphi \colon A \to B$ be a *-homomorphism. Let u in $\mathcal{U}_n(A)$ be given. Use the identification of Proposition 8.1.6 to show that

$$K_1(\varphi)([u]_1) = [\varphi(u) + 1_n - \varphi(1_n)]_1.$$

Conclude that $K_1(\varphi)([u]_1) = [\varphi(u)]_1$ if φ is unital.

Exercise 8.6. Let $\varphi \colon A \to B$ be a surjective *-homomorphism between unital C^*-algebras A and B, and let u be unitary in $\mathcal{U}_n(B)$. Show that $[u]_1$ belongs to $\mathrm{Im}(K_1(\varphi))$ if and only if there exist a natural number $m \geqslant n$ and v in $\mathcal{U}_m(A)$ such that $\varphi(v) = u \oplus 1_{m-n}$.

Exercise 8.7. Let A be an AF-algebra. Show that $K_1(A) = 0$. [Hint: Give a proof of this using Proposition 8.2.7, and try also to prove it directly (without using Proposition 8.2.7), observing that $M_n(\widetilde{A})$ is an AF-algebra for each positive integer n.]

Exercise 8.8. Let A be a C^*-algebra. Recall from Exercise 4.10 the definition of the group of automorphisms, $\mathrm{Aut}(A)$, the group of inner automorphisms, $\mathrm{Inn}(A)$, and the group of approximately inner automorphisms, $\overline{\mathrm{Inn}}(A)$, on A.

(i) Show that $K_1(\alpha) = \mathrm{id}$ for each inner automorphism α.

(ii) Show that $K_1(\alpha) = \mathrm{id}$ for each approximately inner automorphism α.

(iii) Show that $K_1(\alpha)$ is an automorphism on $K_1(A)$ for each automorphism α on A.

(iv) Show that the map $\mathrm{Aut}(A) \to \mathrm{Aut}(K_1(A))$, $\alpha \mapsto K_1(\alpha)$, is a group homomorphism.

(v) Give an example of a C^*-algebra A and an automorphism α on A such that $K_1(\alpha) \neq \mathrm{id}$. [Hint: Try $A = C(\mathbb{T}) \oplus C(\mathbb{T})$.]

Exercise 8.9. Let A be a unital C^*-algebra.

(i) Let u be a unitary element and let s be an isometry in A. Show that $sus^* + (1 - ss^*)$ is unitary and that $[u]_1 = [sus^* + (1 - ss^*)]_1$ (with the identification from Proposition 8.1.6). [Hint: Put

$$v = \begin{pmatrix} s & 1 - ss^* \\ 0 & s^* \end{pmatrix},$$

and calculate $v\,\mathrm{diag}(u, 1)v^*$.]

(ii) Let $u_1, u_2, \ldots, u_n$ be unitary elements in A, and let $s_1, s_2, \ldots, s_n$ be isometries in A such that the range projections $s_1s_1^*, s_2s_2^*, \ldots, s_ns_n^*$ are mutually orthogonal. Show that

$$u = s_1u_1s_1^* + s_2u_2s_2^* + \cdots + s_nu_ns_n^* + (1 - s_1s_1^* - \cdots - s_ns_n^*)$$

is unitary, and that $[u]_1 = [u_1]_1 + [u_2]_1 + \cdots + [u_n]_1$. [Hint: Write u as a suitable product of n unitaries, and use (i).]

(iii) Let $s_1, \ldots, s_n$ be as in question (ii). Put

$$t = \begin{pmatrix} s_1 & s_2 & \cdots & s_n \\ 0 & 0 & \cdots & 0 \\ \vdots & \vdots & \ddots & \vdots \\ 0 & 0 & \cdots & 0 \end{pmatrix}.$$

Show that t is an isometry in $M_n(A)$, and show that for each unitary u in $M_n(A)$ there is a unitary v in A such that $tut^* + (1_n - tt^*) = v \oplus 1_{n-1}$.

(iv) Suppose that A is properly infinite, see Exercise 4.6. Show that

$$K_1(A) = \{[u]_1 : u \in \mathcal{U}(A)\},$$

and conclude that the map $\omega \colon \mathcal{U}(A)/\mathcal{U}_0(A) \to K_1(A)$ is surjective. [Hint: Use (iii) and (i).]

Exercise 8.10. Let $\mathcal{O}_n$ be the Cuntz algebra from Exercise 4.5. Show that $K_1(\mathcal{O}_2) = 0$ and, more generally, that $(n-1)g = 0$ for each $n \geqslant 2$ and for each g in $K_1(\mathcal{O}_n)$.

[Hint: Let $\lambda \colon \mathcal{O}_n \to \mathcal{O}_n$ be as in the proof of Exercise 4.5. Following the lines of that exercise, show that $K_1(\lambda)(g) = g$ and, using Exercise 8.9, that $K_1(\lambda)(g) = ng$ for each g in $K_1(\mathcal{O}_n)$.]

Note. Cuntz showed in [13] that $K_1(\mathcal{O}_n) = 0$ for all n.

Exercise 8.11. Let A be a unital C^*-algebra, let p be a projection in A, $p \neq 1$, and let u_0 be a unitary element in the corner $(1-p)A(1-p)$, i.e., $u_0^* u_0 = 1 - p = u_0 u_0^*$. Put $u = p + u_0$.

(i) Show that u is a unitary element in A.

(ii) Suppose that p is properly infinite and full (see Exercises 4.6, 4.8, and 4.9). Show that $[u]_1 = 0$ if and only if $u \sim_h 1$ in $\mathcal{U}(A)$. [Hint: Assume that $[u]_1 = 0$, find a natural number n and a continuous path $t \mapsto w_t$ in $\mathcal{U}_{n+1}(A)$ with $w_0 = 1_{n+1}$ and $w_1 = u \oplus 1_n$. Use Exercise 4.9 (i) to find v_0 in $M_{1,n+1}(A)$ with $v_0^* v_0 = p \oplus 1_n$ and $v_0 v_0^* \leqslant p$. Put

$$v = \begin{pmatrix} 1-p & 0 & \cdots & 0 \end{pmatrix} + v_0 \in M_{1,n+1}(A),$$

and put $z_t = v w_t v^* + (1 - vv^*)$. Show that $t \mapsto z_t$ is a continuous path of unitary elements in A with $z_0 = 1$ and $z_1 = u$.]

Exercise 8.12. A C^*-algebra A is said to have property (SP) ("small projections") if every non-zero hereditary sub-C^*-algebra of A contains a non-zero projection (see Exercise 5.7).

Show that if A is a unital C^*-algebra with property (SP), then for each unitary element u in A there is a non-zero projection p in A and a unitary element u_0 in the corner $(1 - p)A(1 - p)$ such that $u \sim_h p + u_0$ (see Exercise 8.11).

[Hint: First prove the claim when 1 does not belong to sp(u). Next assume that 1 does belong to sp(u). Take $\varepsilon > 0$ (to be determined later), and choose continuous functions $f, g \colon \mathrm{sp}(u) \to [0,1]$ satisfying: $g(1) = 1$, $f(t)g(t) = g(t)$ for all t in sp(u), and $f(t) = 0$ when $|t - 1| > \varepsilon$. Observe that $g(u)$ is a non-zero, positive element in A, and take a non-zero projection p in $\overline{g(u)Ag(u)}$ (see Exercise 5.7). Put $x = (1 - p)u(1 - p)$.

   (i) Show that $f(u)p = p = pf(u)$, that $\|f(u)u - f(u)\| \leqslant \varepsilon$, and conclude that $\|pu - p\| \leqslant \varepsilon$ and $\|up - p\| \leqslant \varepsilon$.

   (ii) Put $z = pu(1 - p)$. Show that $z = (pu - p)(1 - p)$ and $(1 - p) - x^*x = z^*z$, and conclude that

$$\|z\| \leqslant \varepsilon, \qquad \|u - (p + x)\| \leqslant 2\varepsilon, \qquad \|(1 - p) - x^*x\| \leqslant \varepsilon^2.$$

   (iii) Apply an argument similar to (ii) to show that $\|(1 - p) - xx^*\| \leqslant \varepsilon^2$.

   (iv) Use Exercise 2.8 to show that if ε in $(0, 1/3)$ is chosen small enough, then there is a unitary element u_0 in the corner $(1 - p)A(1 - p)$ such that $\|x - u_0\| \leqslant 1/3$. Conclude that $\|u - (p + u_0)\| < 1$.]

Exercise 8.13. Show that the map

$$\omega \colon \mathcal{U}(A)/\mathcal{U}_0(A) \to K_1(A)$$

is an isomorphism when A is a purely infinite, simple, unital C^*-algebra (see Exercise 5.7). [Hint: Use Exercises 8.9 (iv), 5.7, 8.11, and 8.12.]

Note. There is a spectacular classification result, proved by E. Kirchberg in [28] and by C. Phillips in [32], for the class $\mathcal{C}$ of unital, separable, purely infinite, simple C^*-algebras that are *nuclear* and belong to the bootstrap class $\mathcal{N}$. (See for example Blackadar's book [3] for an explanation of nuclearity and of the bootstrap class $\mathcal{N}$.) The Cuntz algebras $\mathcal{O}_n$ (from Exercise 4.5) belong to $\mathcal{C}$.

The classification theorem says that two C^*-algebras A and B in the class $\mathcal{C}$ are isomorphic if and only if $(K_0(A), [1_A]_0) \cong (K_0(B), [1_B]_0)$ and $K_1(A) \cong K_1(B)$. (We use the convention $(G_0, g_0) \cong (H_0, h_0)$ if there is an isomorphism $\alpha_0 \colon G_0 \to H_0$ with $\alpha_0(g_0) = h_0$.) Moreover, for every pair G_0, G_1 of countable Abelian groups and for every element g_0 in G_0, there is a C^*-algebra A in the class $\mathcal{C}$ such that $(K_0(A), [1_A]_0) \cong (G_0, g_0)$ and $K_1(A) \cong G_1$.

Exercise 8.14. This exercise requires knowledge about von Neumann algebras. Show that $K_1(\mathcal{M}) = 0$ for every von Neumann algebra $\mathcal{M}$. [Hint: Look at Example 8.1.8.]

Exercise 8.15. Show that

$$\mathcal{U}(C(\mathbb{T}^n))/\mathcal{U}_0(C(\mathbb{T}^n)) = \pi^1(\mathbb{T}^n) \cong \mathbb{Z}^n$$

for each n in $\mathbb{N}$. [Hint: Introduce, as in Example 8.3.2, a winding number $w \colon C(\mathbb{T}^n, \mathbb{T}) \to \mathbb{Z}^n$, and show that it induces an isomorphism $\pi^1(\mathbb{T}^n) \to \mathbb{Z}^n$ given by $[u] \mapsto w(u)$.]

Exercise 8.16. Prove that K_1 is split exact (Proposition 8.2.5) by following the lines of the proof of Proposition 4.3.3.

Exercise 8.17. Let A be a unital C^*-algebra. Show that

$$(8.9) \qquad \mathcal{U}(\widetilde{KA})/\mathcal{U}_0(\widetilde{KA}) \cong \mathcal{U}_\infty(A)/\sim_1 \cong K_1(A).$$

[Hint: Let $\kappa_n \colon M_n(A) \to KA$ be as in Section 6.4, put $e_n = \kappa_n(1_n)$, and define $\rho_n \colon \mathcal{U}_n(A) \to \mathcal{U}(\widetilde{KA})$ by $\rho_n(u) = \kappa_n(u) + (1 - e_n)$. Check that $\rho_{n+k}(u \oplus 1_k) = \rho_n(u)$ for every u in $\mathcal{U}_n(A)$. Let v be a unitary element in $\widetilde{KA}$ with $s(v) = 1$. Show that

$$\lim_{n \to \infty} \|v - (e_n v e_n + (1 - e_n))\| = 0,$$

and use this, and Exercise 2.8, to show that $v \sim_h \rho_n(u)$ for some n in $\mathbb{N}$ and some u in $\mathcal{U}_n(A)$. Combine these facts to establish (8.9).]

Exercise 8.18. Let A be a unital C^*-algebra, and consider an element

$$a = \begin{pmatrix} a_{11} & a_{12} \\ a_{21} & a_{22} \end{pmatrix} \in M_2(A).$$

(i) Suppose that a_{22} is invertible. Show that there are elements x, y in A such that

$$a = \begin{pmatrix} 1 & x \\ 0 & 1 \end{pmatrix} \begin{pmatrix} a_{11} - a_{12}a_{22}^{-1}a_{21} & 0 \\ 0 & a_{22} \end{pmatrix} \begin{pmatrix} 1 & 0 \\ y & 1 \end{pmatrix}.$$

(ii) Suppose again that a_{22} is invertible. Show that a is invertible if and only if $a_{11} - a_{12}a_{22}^{-1}a_{21}$ is invertible.

(iii) Suppose that a_{22} and a are invertible. Show that there is an invertible element b in A such that $[a]_1 = [b]_1$ in $K_1(A)$ (using the identification in Remark 8.1.7).

(iv) Suppose that $\mathrm{sr}(A) = 1$, i.e., that $GL(A)$ is a dense subset of A. Let c be an invertible element in $M_2(A)$. Show that there is an invertible element b in A with $[b]_1 = [a]_1$ in $K_1(A)$. [Hint: First find $a = (a_{ij})$ in $M_2(A)$ with a_{22} invertible and such that $\|a - c\|$ is so small that c is invertible and $a \sim_h c$ in $GL_2(A)$.]

Note: Rieffel proved in [37] that if $\mathrm{sr}(A) = 1$, then $\mathrm{sr}(M_n(A)) = 1$ for all natural numbers n, and in this case the map $\omega \colon \mathcal{U}(A) \to K_1(A)$ is surjective.

Chapter 9

The Index Map

In this chapter we introduce the index map associated to a short exact sequence

$$(9.1) \qquad 0 \longrightarrow I \xrightarrow{\;\varphi\;} A \xrightarrow{\;\psi\;} B \longrightarrow 0$$

of C^*-algebras. The index map is a group homomorphism $\delta_1 \colon K_1(B) \to K_0(I)$ that gives rise to an exact sequence

$$(9.2)$$

$$
\begin{array}{ccc}
K_1(I) \xrightarrow{\;K_1(\varphi)\;} K_1(A) \xrightarrow{\;K_1(\psi)\;} K_1(B) \\
\Big\downarrow {\scriptstyle \delta_1} \\
K_0(B) \xleftarrow{\;K_0(\psi)\;} K_0(A) \xleftarrow{\;K_0(\varphi)\;} K_0(I).
\end{array}
$$

The index map measures an obstruction to lifting unitaries in (a matrix algebra over) B (with a unit adjoined) to a unitary in a matrix algebra over A (with a unit adjoined), see Exercise 9.2. The index map generalizes the classical Fredholm index of Fredholm operators on a Hilbert space (as shown in Section 9.4). The existence of the index map makes it possible to calculate — even without knowing what the map looks like — the K-groups of some C^*-algebras by fitting them into suitable short exact sequences. As an example of this technique we obtain in Exercise 9.5 that $K_0(C_0(\mathbb{R}^2))$ is isomorphic to $K_1(C(\mathbb{T}))$.

9.1   Definition of the index map

The definition of the index map is based on the following two lemmas.

Lemma 9.1.1. *Suppose we are given the short exact sequence (9.1) of C^*-algebras, and let u in $\mathcal{U}_n(\widetilde{B})$ be given.*

(i) *There exist a unitary v in $\mathcal{U}_{2n}(\widetilde{A})$ and a projection p in $\mathcal{P}_{2n}(\widetilde{I})$ such that*

$$\widetilde{\psi}(v) = \begin{pmatrix} u & 0 \\ 0 & u^* \end{pmatrix}, \qquad \widetilde{\varphi}(p) = v \begin{pmatrix} 1_n & 0 \\ 0 & 0 \end{pmatrix} v^*, \qquad s(p) = \begin{pmatrix} 1_n & 0 \\ 0 & 0 \end{pmatrix}.$$

(ii) *If v and p are as in (i), and if w in $\mathcal{U}_{2n}(\widetilde{A})$ and q in $\mathcal{P}_{2n}(\widetilde{I})$ satisfy*

$$\widetilde{\psi}(w) = \begin{pmatrix} u & 0 \\ 0 & u^* \end{pmatrix}, \qquad \widetilde{\varphi}(q) = w \begin{pmatrix} 1_n & 0 \\ 0 & 0 \end{pmatrix} w^*,$$

then $s(q) = \mathrm{diag}(1_n, 0_n)$ and $p \sim_u q$ in $\mathcal{P}_{2n}(\widetilde{I})$.

Proof. (i). Apply Lemma 2.1.7 to obtain a unitary v in $\mathcal{U}_{2n}(\widetilde{A})$ with $\widetilde{\psi}(v) = \mathrm{diag}(u, u^*)$, so that

$$\widetilde{\psi}\left(v \begin{pmatrix} 1_n & 0 \\ 0 & 0 \end{pmatrix} v^* \right) = \begin{pmatrix} 1_n & 0 \\ 0 & 0 \end{pmatrix}.$$

According to Lemma 4.3.1, $\widetilde{\varphi}$ is injective and there exists p in $M_{2n}(\widetilde{I})$ such that $\widetilde{\varphi}(p) = v\,\mathrm{diag}(1_n, 0)v^*$. The element p is necessarily a projection. Since

$$\widetilde{\psi}(\widetilde{\varphi}(p)) = \widetilde{\psi}\left(v \begin{pmatrix} 1_n & 0 \\ 0 & 0 \end{pmatrix} v^* \right) = \begin{pmatrix} 1_n & 0 \\ 0 & 0 \end{pmatrix},$$

it follows that $s(p) = \mathrm{diag}(1_n, 0)$.

(ii). The argument in (i) used to show that $s(p) = \mathrm{diag}(1_n, 0_n)$ also shows that $s(q) = \mathrm{diag}(1_n, 0_n)$. Note that $\widetilde{\psi}(wv^*) = 1_{2n}$. By Lemma 4.3.1 there is an element z in $M_{2n}(\widetilde{I})$ such that $\widetilde{\varphi}(z) = wv^*$, and z is necessarily unitary because $\widetilde{\varphi}$ is injective. Since $\widetilde{\varphi}(zpz^*) = \widetilde{\varphi}(q)$, we infer that $q = zpz^*$. Thus $p \sim_u q$ in $\mathcal{P}_{2n}(\widetilde{I})$ as desired. $\square$

Define $\nu \colon \mathcal{U}_\infty(\widetilde{B}) \to K_0(I)$ by $\nu(u) = [p]_0 - [s(p)]_0$ for u in $\mathcal{U}_n(\widetilde{A})$, where p in $\mathcal{P}_{2n}(\widetilde{I})$ corresponds to u as in Lemma 9.1.1 (i). The map ν is well-defined by Lemma 9.1.1 (ii).

Lemma 9.1.2. *The map* $\nu\colon \mathcal{U}_\infty(\widetilde{B}) \to K_0(I)$ *has the following properties:*

(i) $\nu(u_1 \oplus u_2) = \nu(u_1) + \nu(u_2)$ *for all* u_1, u_2 *in* $\mathcal{U}_\infty(\widetilde{B})$,

(ii) $\nu(1) = 0$,

(iii) *if* u_1, u_2 *belong to* $\mathcal{U}_n(\widetilde{B})$ *and* $u_1 \sim_h u_2$, *then* $\nu(u_1) = \nu(u_2)$,

(iv) $\nu(\widetilde{\psi}(u)) = 0$ *for every* u *in* $\mathcal{U}_\infty(\widetilde{A})$,

(v) $K_0(\varphi)(\nu(u)) = 0$ *for every* u *in* $\mathcal{U}_\infty(\widetilde{B})$.

Proof. (i). For $j = 1, 2$, let u_j in $\mathcal{U}_{n_j}(\widetilde{B})$ be given, and choose v_j in $\mathcal{U}_{2n_j}(\widetilde{A})$ and p_j in $\mathcal{P}_{2n_j}(\widetilde{I})$ satisfying (i) in Lemma 9.1.1, i.e.,

$$\widetilde{\psi}(v_j) = \begin{pmatrix} u_j & 0 \\ 0 & u_j^* \end{pmatrix}, \qquad \widetilde{\varphi}(p_j) = v_j \begin{pmatrix} 1_{n_j} & 0 \\ 0 & 0 \end{pmatrix} v_j^*,$$

so that $\nu(u_j) = [p_j]_0 - [s(p_j)]_0$. Introduce elements y in $\mathcal{U}_{2(n_1+n_2)}(\mathbb{C})$, v in $\mathcal{U}_{2(n_1+n_2)}(\widetilde{A})$, and p in $\mathcal{U}_{2(n_1+n_2)}(\widetilde{I})$ by

$$y = \begin{pmatrix} 1_{n_1} & 0 & 0 & 0 \\ 0 & 0 & 1_{n_2} & 0 \\ 0 & 1_{n_1} & 0 & 0 \\ 0 & 0 & 0 & 1_{n_2} \end{pmatrix}, \qquad v = y \begin{pmatrix} v_1 & 0 \\ 0 & v_2 \end{pmatrix} y^*, \qquad p = y \begin{pmatrix} p_1 & 0 \\ 0 & p_2 \end{pmatrix} y^*.$$

Then

$$\widetilde{\psi}(v) = \begin{pmatrix} u_1 & 0 & 0 & 0 \\ 0 & u_2 & 0 & 0 \\ 0 & 0 & u_1^* & 0 \\ 0 & 0 & 0 & u_2^* \end{pmatrix}, \qquad \widetilde{\varphi}(p) = v \begin{pmatrix} 1_{n_1+n_2} & 0 \\ 0 & 0 \end{pmatrix} v^*,$$

and so

$$\nu(u_1 \oplus u_2) = [p]_0 - [s(p)]_0 = [p_1 \oplus p_2]_0 - [s(p_1 \oplus p_2)]_0 = \nu(u_1) + \nu(u_2)$$

because $p \sim_u p_1 \oplus p_2$.

(iii). Choose v_1 in $\mathcal{U}_{2n}(\widetilde{A})$ and p_1 in $\mathcal{P}_{2n}(\widetilde{I})$ such that

$$\widetilde{\psi}(v_1) = \begin{pmatrix} u_1 & 0 \\ 0 & u_1^* \end{pmatrix}, \qquad \widetilde{\varphi}(p_1) = v_1 \begin{pmatrix} 1_n & 0 \\ 0 & 0 \end{pmatrix} v_1^*.$$

Then $\nu(u_1) = [p_1]_0 - [s(p_1)]_0$. Since $u_1^* u_2 \sim_h 1_n \sim_h u_1 u_2^*$, we can apply Lemma 2.1.7 to get unitary elements a, b in $M_n(\widetilde{A})$ with $\psi(a) = u_1^* u_2$ and

$\widetilde{\psi}(b) = u_1 u_2^*$. Putting $v_2 = v_1 \operatorname{diag}(a,b)$ in $\mathcal{U}_{2n}(\widetilde{A})$ we obtain

$$\widetilde{\psi}(v_2) = \begin{pmatrix} u_2 & 0 \\ 0 & u_2^* \end{pmatrix}, \qquad v_2 \begin{pmatrix} 1_n & 0 \\ 0 & 0 \end{pmatrix} v_2^* = v_1 \begin{pmatrix} 1_n & 0 \\ 0 & 0 \end{pmatrix} v_1^* = \widetilde{\varphi}(p_1).$$

By the definition of ν we conclude that $\nu(u_2) = [p_1]_0 - [s(p_1)]_0 = \nu(u_1)$.

(iv). Put $v = \operatorname{diag}(u, u^*)$ in $\mathcal{U}_{2n}(\widetilde{A})$, and put $p = \operatorname{diag}(1_n, 0)$ in $\mathcal{P}_{2n}(\widetilde{I})$, so that $p = s(p)$. Then

$$\widetilde{\psi}(v) = \begin{pmatrix} \widetilde{\psi}(u) & 0 \\ 0 & \widetilde{\psi}(u^*) \end{pmatrix}, \qquad \widetilde{\varphi}(p) = v \begin{pmatrix} 1_n & 0 \\ 0 & 0 \end{pmatrix} v^*.$$

Hence $\nu(\widetilde{\psi}(u)) = [p]_0 - [s(p)]_0 = 0$.

Part (ii) is an immediate consequence of (iv), and (v) follows from the fact that $\widetilde{\varphi}(p)$ is unitarily equivalent to $s(\widetilde{\varphi}(p))$ in $M_{2n}(\widetilde{A})$, when p is a projection in $M_{2n}(\widetilde{I})$ associated to u as in Lemma 9.1.1 (i).   □

Definition 9.1.3 (The index map). Suppose we are given the short exact sequence (9.1). As above, let $\nu \colon \mathcal{U}_\infty(\widetilde{B}) \to K_0(I)$ be the map given by $\nu(u) = [p]_0 - [s(p)]_0$, when u belongs to $\mathcal{U}_n(\widetilde{B})$ and p in $\mathcal{P}_{2n}(\widetilde{I})$ corresponds to u as in Lemma 9.1.1 (i). By the universal property of K_1 (Proposition 8.1.5) and by Lemma 9.1.2, there is a unique group homomorphism $\delta_1 \colon K_1(B) \to K_0(I)$ satisfying $\delta_1([u]_1) = \nu(u)$ for each u in $\mathcal{U}_\infty(\widetilde{B})$.

The map δ_1 is called the *index map* associated with the short exact sequence (9.1).

We summarize the basic properties of the index map in the following standard picture.

Proposition 9.1.4 (First standard picture of the index map).   *Let*

$$0 \longrightarrow I \overset{\varphi}{\longrightarrow} A \overset{\psi}{\longrightarrow} B \longrightarrow 0$$

be a short exact sequence of C^*-algebras. Let n be a natural number, and suppose that u in $\mathcal{U}_n(\widetilde{B})$, v in $\mathcal{U}_{2n}(\widetilde{A})$, and p in $\mathcal{P}_{2n}(\widetilde{I})$ satisfy

$$\widetilde{\varphi}(p) = v \begin{pmatrix} 1_n & 0 \\ 0 & 0 \end{pmatrix} v^*, \qquad \widetilde{\psi}(v) = \begin{pmatrix} u & 0 \\ 0 & u^* \end{pmatrix}.$$

Then $\delta_1([u]_1) = [p]_0 - [s(p)]_0$. Moreover,

(i) $\delta_1 \circ K_1(\psi) = 0$,

(ii) $K_0(\varphi) \circ \delta_1 = 0$.

Proposition 9.1.5 (Naturality of the index map). *Let*

(9.3)
$$
\begin{array}{ccccccccc}
0 & \longrightarrow & I & \overset{\varphi}{\longrightarrow} & A & \overset{\psi}{\longrightarrow} & B & \longrightarrow & 0 \\
& & \gamma\downarrow & & \alpha\downarrow & & \downarrow\beta & & \\
0 & \longrightarrow & I' & \underset{\varphi'}{\longrightarrow} & A' & \underset{\psi'}{\longrightarrow} & B' & \longrightarrow & 0
\end{array}
$$

be a commutative diagram with short exact rows of C^*-algebras, and where α, β, and γ are *-homomorphisms. Let $\delta_1 \colon K_1(B) \to K_0(I)$ and $\delta_1' \colon K_1(B') \to K_0(I')$ be the index maps associated with the short exact sequences in the upper and lower rows of (9.3), respectively. Then the diagram

$$
\begin{array}{ccc}
K_1(B) & \overset{\delta_1}{\longrightarrow} & K_0(I) \\
K_1(\beta)\downarrow & & \downarrow K_0(\gamma) \\
K_1(B') & \underset{\delta_1'}{\longrightarrow} & K_0(I')
\end{array}
$$

is commutative.

Proof. Let g be an element in $K_1(B)$. Find a natural number n and a unitary u in $\mathcal{U}_n(\widetilde{B})$ with $g = [u]_1$. Apply Lemma 9.1.1 (i) to get a unitary v in $\mathcal{U}_{2n}(\widetilde{A})$ and a projection p in $\mathcal{P}_{2n}(\widetilde{I})$ such that

$$
\widetilde{\psi}(v) = \begin{pmatrix} u & 0 \\ 0 & u^* \end{pmatrix}, \qquad \widetilde{\varphi}(p) = v \begin{pmatrix} 1_n & 0 \\ 0 & 0 \end{pmatrix} v^*.
$$

Put $v' = \widetilde{\alpha}(v)$ in $\mathcal{U}_{2n}(\widetilde{A'})$ and $p' = \widetilde{\gamma}(p)$ in $\mathcal{P}_{2n}(\widetilde{I'})$. Then

$$
\widetilde{\psi'}(v') = (\psi' \circ \alpha)^\sim(v) = (\beta \circ \psi)^\sim(v) = \widetilde{\beta}(\widetilde{\psi}(v)) = \begin{pmatrix} \widetilde{\beta}(u) & 0 \\ 0 & \widetilde{\beta}(u)^* \end{pmatrix},
$$

$$
\widetilde{\varphi'}(p') = (\varphi' \circ \gamma)^\sim(p) = (\alpha \circ \varphi)^\sim(p) = \widetilde{\alpha}(v) \begin{pmatrix} 1_n & 0 \\ 0 & 0 \end{pmatrix} \widetilde{\alpha}(v)^*
$$

$$
= v' \begin{pmatrix} 1_n & 0 \\ 0 & 0 \end{pmatrix} (v')^*.
$$

From the definition of the index map, we have

$$(\delta_1' \circ K_1(\beta))(g) = \delta_1'([\widetilde{\beta}(u)]_1) = [p']_0 - [s(p')]_0$$
$$= [\widetilde{\gamma}(p)]_0 - [s(\widetilde{\gamma}(p))]_0 = K_0(\gamma)([p]_0 - [s(p)]_0)$$
$$= K_0(\gamma)(\delta_1([u]_1)) = (K_0(\gamma) \circ \delta_1)(g).$$

This shows that $\delta_1' \circ K_1(\beta) = K_0(\gamma) \circ \delta_1$. $\square$

9.2   The index map and partial isometries

We give in this section another picture of the index map pointed out to us by George Elliott. Besides having a stronger flavor of "index", this picture is more intuitive and in most cases easier to work with than the picture of the index map described in the previous section.

The key observation is as follows.

Lemma 9.2.1. *Let* $\psi \colon A \to B$ *be a surjective* *-*homomorphism between* C^*-*algebras* A *and* B, *and suppose that* A *is unital, in which case also* B *is unital and* ψ *is unit preserving. Then for each unitary element* u *in* B *there is a partial isometry* v *in* $M_2(A)$ *such that*

(9.4) $$\psi(v) = \begin{pmatrix} u & 0 \\ 0 & 0 \end{pmatrix}.$$

Proof. Take a lift a in A of u with $\|a\| = 1$ (see Paragraph 2.2.10), and put

$$v = \begin{pmatrix} a & 0 \\ (1 - a^*a)^{1/2} & 0 \end{pmatrix}.$$

Then $v^*v = \operatorname{diag}(1, 0)$, which entails that v is a partial isometry. Use the identity

$$\psi((1 - a^*a)^{1/2}) = (1 - u^*u)^{1/2} = 0$$

to see that (9.4) holds. $\square$

Proposition 9.2.2 (Second standard picture of the index map). *Let*

$$0 \longrightarrow I \overset{\varphi}{\longrightarrow} A \overset{\psi}{\longrightarrow} B \longrightarrow 0$$

be a short exact sequence of C^*-*algebras. Let* $n \leqslant m$ *be natural numbers, let*

u *be a unitary element in* $\mathcal{U}_n(\widetilde{B})$, *and let* v *be a partial isometry in* $M_m(\widetilde{A})$ *with*

$$(9.5) \qquad \widetilde{\psi}(v) = \begin{pmatrix} u & 0 \\ 0 & 0_{m-n} \end{pmatrix}.$$

Then $1_m - v^*v = \widetilde{\varphi}(p)$ *and* $1_m - vv^* = \widetilde{\varphi}(q)$ *for some projections* p, q *in* $\mathcal{P}_m(\widetilde{I})$, *and the index map* $\delta_1 \colon K_1(B) \to K_0(I)$ *is given by*

$$(9.6) \qquad \delta_1([u]_1) = [p]_0 - [q]_0.$$

Lemma 9.2.1 shows that we can always find a partial isometry v that satisfies (9.5). It is a consequence of Proposition 9.2.2 that the element on the right-hand side of (9.6) does not depend on the choice of v. It is also a consequence of Proposition 9.2.2 that this element actually belongs to $K_0(I)$. (A priori, we only know that it belongs to $K_0(\widetilde{I})$.)

Proof of proposition. Set $e = 1_m - v^*v$ and $f = 1_m - vv^*$ in $\mathcal{P}_m(\widetilde{A})$. Then $\widetilde{\psi}(e) = \widetilde{\psi}(f) = \mathrm{diag}(0_n, 1_{m-n})$. Because $\widetilde{\psi}(e)$ and $\widetilde{\psi}(f)$ are scalar matrices there are projections p and q in $\mathcal{P}_m(\widetilde{I})$ such that $\widetilde{\varphi}(p) = e$ and $\widetilde{\varphi}(q) = f$ (see Lemma 4.3.1) and $s(p) = s(q) = \mathrm{diag}(0_n, 1_{m-n})$. Put

$$w = \begin{pmatrix} v & f \\ e & v^* \end{pmatrix}, \qquad r = \begin{pmatrix} 1_m - q & 0 \\ 0 & p \end{pmatrix}, \qquad z = \begin{pmatrix} 1_n & 0 & 0_n & 0 \\ 0 & 0_{m-n} & 0 & 1_{m-n} \\ 0_n & 0 & 1_n & 0 \\ 0 & 1_{m-n} & 0 & 0_{m-n} \end{pmatrix}.$$

Then r is a projection in $M_{2m}(\widetilde{I})$, w is a unitary element in $M_{2m}(\widetilde{A})$, z is a self-adjoint unitary matrix in $M_{2m}(\mathbb{C})$, and

$$\widetilde{\psi}(zw) = z \begin{pmatrix} u & 0 & 0_n & 0 \\ 0 & 0_{m-n} & 0 & 1_{m-n} \\ 0_n & 0 & u^* & 0 \\ 0 & 1_{m-n} & 0 & 0_{m-n} \end{pmatrix} = \begin{pmatrix} u_1 & 0 \\ 0 & u_1^* \end{pmatrix},$$

where $u_1 = \mathrm{diag}(u, 1_{m-n})$ in $\mathcal{U}_m(\widetilde{B})$. Now

$$zw \begin{pmatrix} 1_m & 0 \\ 0 & 0 \end{pmatrix} w^* z^* = z \begin{pmatrix} vv^* & ve \\ ev^* & e \end{pmatrix} z^* = z \begin{pmatrix} 1_m - f & 0 \\ 0 & e \end{pmatrix} z^* = \widetilde{\varphi}(zrz^*).$$

It follows from the definition of the index map that

$$\delta_1([u]_1) = \delta_1([u_1]_1) = [zrz^*]_0 - [s(zrz^*)]_0 = [r]_0 - [s(r)]_0$$
$$= [1_m - q]_0 + [p]_0 - [1_n]_0 - [1_{m-n}]_0 = [p]_0 - [q]_0,$$

as desired.   □

If I is an ideal in A and φ is the inclusion mapping in Proposition 9.2.2, then (9.6) can be rephrased as

(9.7) $$\delta_1([u]_1) = [1_m - v^*v]_0 - [1_m - vv^*]_0,$$

where m and n are integers with $m \geqslant n$, u belongs to $\mathcal{U}_n(\widetilde{B})$, and v is a partial isometry in $M_m(\widetilde{A})$ that lifts $\text{diag}(u, 0_{m-n})$.

It is convenient to have an expression for the index map that does not involve unitizations when A and B are unital C^*-algebras. This is given in the following proposition, where both pictures of the index map are described.

Proposition 9.2.3. *Let*

$$0 \longrightarrow I \overset{\varphi}{\longrightarrow} A \overset{\psi}{\longrightarrow} B \longrightarrow 0$$

*be a short exact sequence of C^*-algebras, and suppose that A is unital, in which case B is unital, too, and ψ is unit preserving. Let $\overline{\varphi} \colon \widetilde{I} \to A$ be the *-homomorphism given by $\overline{\varphi}(x + \alpha 1_{\widetilde{I}}) = \varphi(x) + \alpha 1_A$ for x in I and α in $\mathbb{C}$. Let u be a unitary element in $M_n(B)$.*

(i) *If v is a unitary element in $M_{2n}(A)$ and p is a projection in $M_{2n}(\widetilde{I})$ such that*

$$\overline{\varphi}(p) = v \begin{pmatrix} 1_n & 0 \\ 0 & 0 \end{pmatrix} v^*, \qquad \psi(v) = \begin{pmatrix} u & 0 \\ 0 & u^* \end{pmatrix},$$

   then $\delta_1([u]_1) = [p]_0 - [s(p)]_0$.

(ii) *If $m \geqslant n$ is an integer and v is a partial isometry in $M_m(A)$ with $\psi(v) = \text{diag}(u, 0_{m-n})$, then $1_m - v^*v = \overline{\varphi}(p)$ and $1_m - vv^* = \overline{\varphi}(q)$ for some projections p, q in $M_m(\widetilde{I})$, and $\delta_1([u]_1) = [p]_0 - [q]_0$.*

Proof. (i). For each natural number k, set

$$f_k = 1_{M_k(\widetilde{A})} - 1_{M_k(A)}, \qquad g_k = 1_{M_k(\widetilde{B})} - 1_{M_k(B)}.$$

Then $u_1 = u + g_n$ is a unitary element in $M_n(\widetilde{B})$, $v_1 = v + f_{2n}$ is a unitary element in $M_{2n}(\widetilde{A})$, and $\widetilde{\psi}(v_1) = \mathrm{diag}(u_1, u_1^*)$. We wish to determine the scalar part $s(p)$ of p. Observe to this end that $\psi(\overline{\varphi}(s(p))) = \psi(\overline{\varphi}(p)) = \mathrm{diag}(1_{M_n(B)}, 0_n)$. We obtain therefore

$$s(p) = \begin{pmatrix} 1_{M_n(\widetilde{I})} & 0 \\ 0 & 0 \end{pmatrix}, \qquad \overline{\varphi}(s(p)) = \begin{pmatrix} 1_{M_n(A)} & 0 \\ 0 & 0 \end{pmatrix}.$$

Hence

$$\widetilde{\varphi}(p) = \varphi(p - s(p)) + \widetilde{\varphi}(s(p)) = \overline{\varphi}(p) - \overline{\varphi}(s(p)) + \widetilde{\varphi}(s(p))$$

$$= \overline{\varphi}(p) + \begin{pmatrix} f_n & 0 \\ 0 & 0 \end{pmatrix} = v_1 \begin{pmatrix} 1_{M_n(\widetilde{A})} & 0 \\ 0 & 0 \end{pmatrix} v_1^*.$$

From the first standard picture of the index map (Proposition 9.1.4) we conclude that $\delta_1([u_1]_1) = [p]_0 - [s(p)]_0$, and the identification $[u]_1 = [u_1]_1$ from Proposition 8.1.6 completes the proof.

(ii). We assume in the proof that $\varphi(I) \neq A$ — an assumption that entails that $\overline{\varphi} \colon M_m(\widetilde{I}) \to M_m(A)$ is injective for each natural number m. We leave it to the reader to consider the case where $\varphi(I) = A$.

The image of $\overline{\varphi} \colon M_m(\widetilde{I}) \to M_m(A)$ consists of those elements x for which $\psi(x)$ belongs to $M_m(\mathbb{C}1_B)$. Hence

$$1_{M_m(A)} - v^*v = \overline{\varphi}(p), \qquad 1_{M_m(A)} - vv^* = \overline{\varphi}(q)$$

for some elements p, q in $M_m(\widetilde{I})$ that necessarily are projections because $\overline{\varphi}$ is injective. Since $\psi(\overline{\varphi}(p)) = \mathrm{diag}(0_n, 1_{M_{m-n}(B)})$, we see that

$$s(p) = \begin{pmatrix} 0_n & 0 \\ 0 & 1_{M_{m-n}(\widetilde{I})} \end{pmatrix}.$$

Let f_n and u_1 be as in the proof of (i) above, and put

$$w = v + \begin{pmatrix} f_n & 0 \\ 0 & 0_{m-n} \end{pmatrix},$$

so that w is a partial isometry in $M_m(\widetilde{A})$ with $\widetilde{\psi}(w) = \mathrm{diag}(u_1, 0_{m-n})$. As in (i),

$$\widetilde{\varphi}(p) = \overline{\varphi}(p) - \overline{\varphi}(s(p)) + \widetilde{\varphi}(s(p))$$

$$= 1_{M_m(A)} - v^*v + \begin{pmatrix} 0_n & 0 \\ 0 & f_{m-n} \end{pmatrix} = 1_{M_m(\widetilde{A})} - w^*w,$$

and, similarly, $\widetilde{\varphi}(q) = 1_{M_m(\widetilde{A})} - ww^*$. By the second standard picture of the index map (Proposition 9.2.2) we find that $\delta_1([u_1]_1) = [p]_0 - [q]_0$, which together with the identification $[u]_1 = [u_1]_1$ completes the proof. $\square$

We note the following simplifications of Propositions 9.2.2 and 9.2.3 when u in $\mathcal{U}_n(\widetilde{B})$ or in $\mathcal{U}_n(B)$ lifts to a partial isometry in $M_n(\widetilde{A})$ or in $M_n(A)$, respectively, and where it for further simplification is assumed that I is an ideal in A.

Proposition 9.2.4. *Let*

$$0 \longrightarrow I \stackrel{\iota}{\hookrightarrow} A \stackrel{\psi}{\longrightarrow} B \longrightarrow 0$$

be a short exact sequence of C^*-algebras, where I is an ideal in A and where ι is the inclusion mapping.

(i) *Let u be a unitary element in $M_n(\widetilde{B})$ which has a lift to a partial isometry v in $M_n(\widetilde{A})$, i.e., $\widetilde{\psi}(v) = u$. Then $1_n - v^*v$ and $1_n - vv^*$ are projections in $M_n(I)$, and*

(9.8) $$\delta_1([u]_1) = [1_n - v^*v]_0 - [1_n - vv^*]_0.$$

(ii) *Suppose that A is unital (in which case also B is unital and ψ is unit preserving). Let u be a unitary element in $M_n(B)$ which has a lift to a partial isometry v in $M_n(A)$. Then $1_n - v^*v$ and $1_n - vv^*$ are projections in $M_n(I)$, and*

(9.9) $$\delta_1([u]_1) = [1_n - v^*v]_0 - [1_n - vv^*]_0.$$

Proof. (i). Since

$$\widetilde{\psi}(1_n - v^*v) = 1_n - u^*u = 0, \qquad \widetilde{\psi}(1_n - vv^*) = 1_n - uu^* = 0,$$

we see that $1_n - v^*v$ and $1_n - vv^*$ belong to $M_n(I)$, and these two elements are projections because v is a partial isometry. The identity (9.8) follows from Proposition 9.2.2.

(ii). Here, $\psi(1_n - v^*v) = 1_n - u^*u = 0$, showing that $1_n - v^*v$ belongs to $M_n(I)$, and in a similar way we see that $1_n - vv^*$ belongs to $M_n(I)$. The identity (9.9) follows from Proposition 9.2.3 (ii). □

9.3 An exact sequence of K-groups

We show in this section that each short exact sequence of C^*-algebras

$$0 \longrightarrow I \overset{\varphi}{\longrightarrow} A \overset{\psi}{\longrightarrow} B \longrightarrow 0$$

induces an exact sequence of K-groups (see the diagram (9.2)), with the index map $\delta_1\colon K_1(B) \to K_0(I)$ being the connection between K_1 and K_0.

For ease of notation (and without loss of generality, see Paragraph 1.1.5) we assume in (the proofs of) the following two lemmas that I is an ideal in A and that φ is the inclusion mapping. In this case $M_n(\widetilde{I})$ is a unital sub-C^*-algebra of $M_n(\widetilde{A})$ for each n.

Lemma 9.3.1. *The kernel of the index map $\delta_1\colon K_1(B) \to K_0(I)$ is contained in the image of the map $K_1(\psi)\colon K_1(A) \to K_1(B)$.*

Proof. Suppose that g in $K_1(B)$ belongs to the kernel of δ_1. Take u in $\mathcal{U}_n(\widetilde{B})$ with $g = [u]_1$, and use Lemma 9.2.1 to find a partial isometry w_1 in $M_{2n}(\widetilde{A})$ with

$$\widetilde{\psi}(w_1) = \begin{pmatrix} u & 0 \\ 0 & 0_n \end{pmatrix}.$$

By the second standard picture of the index map (Proposition 9.2.2),

$$0 = \delta_1([u]_1) = [1_{2n} - w_1^*w_1]_0 - [1_{2n} - w_1w_1^*]_0 \quad \text{in} \quad K_0(I).$$

It follows from Proposition 3.1.7 that

$$w_2^*w_2 = (1_{2n} - w_1^*w_1) \oplus 1_k, \qquad w_2w_2^* = (1_{2n} - w_1w_1^*) \oplus 1_k$$

for some natural number k and some partial isometry w_2 in $M_m(\widetilde{I})$, where $m = 2n + k$. Now,

$$\widetilde{\psi}(w_2^*w_2) = \widetilde{\psi}(w_2w_2^*) = \begin{pmatrix} 0_n & 0 \\ 0 & 1_{m-n} \end{pmatrix},$$

and $\widetilde{\psi}(w_2)$ is a scalar matrix because w_2 belongs to $M_n(\widetilde{I})$. This shows that $\widetilde{\psi}(w_2) = \operatorname{diag}(0_n, z)$ for some scalar unitary matrix z in $M_{m-n}(\widetilde{B})$. Having finite spectrum, z is homotopic to 1_{m-n} in $\mathcal{U}_{m-n}(\widetilde{B})$, see Lemma 2.1.3. (With a small amount of extra care one could have chosen w_2 such that $z = 1_{m-n}$.) Set $v = \operatorname{diag}(w_1, 0_k) + w_2$. Then v is a unitary element in $M_m(\widetilde{A})$ (see Exercise 2.6), and

$$\widetilde{\psi}(v) = \begin{pmatrix} u & 0 \\ 0 & 0_{m-n} \end{pmatrix} + \begin{pmatrix} 0_n & 0 \\ 0 & z \end{pmatrix} \sim_h \begin{pmatrix} u & 0 \\ 0 & 1_{m-n} \end{pmatrix} \quad \text{in} \quad \mathcal{U}_m(\widetilde{B}).$$

This proves that

$$g = [u]_1 = [\widetilde{\psi}(v)]_1 = K_1(\psi)([v]_1),$$

as desired. $\square$

Lemma 9.3.2. *The kernel of the map $K_0(\varphi)\colon K_0(I) \to K_0(A)$ is contained in the image of the index map $\delta_1\colon K_1(B) \to K_0(I)$.*

Proof. Let g in $K_0(I)$ be in the kernel of $K_0(\varphi)$. By the appendix to the standard picture for K_0 (Lemma 4.2.3) there are a natural number n, a projection p in $M_n(\widetilde{I})$, and a unitary w in $M_n(\widetilde{A})$ with $g = [p]_0 - [s(p)]_0$ and $wpw^* = s(p)$.

The element $u_0 = \widetilde{\psi}(w(1_n - p))$ is a partial isometry in $M_n(\widetilde{B})$ and

$$1_n - u_0^* u_0 = \widetilde{\psi}(p) = \widetilde{\psi}(s(p)) = 1_n - u_0 u_0^*.$$

It follows that u_0 is a partial unitary element (see Paragraph 8.2.10) and that $u = u_0 + (1_n - u_0^* u_0)$ is a unitary element in $M_n(\widetilde{B})$. We wish to lift $\operatorname{diag}(u, 0_n)$ to a suitable partial isometry v in $M_{2n}(\widetilde{A})$. As a first step in this direction observe that the partial isometry $v_1 = \operatorname{diag}(w(1_n - p), s(p))$ in $M_{2n}(\widetilde{A})$ satisfies $\widetilde{\psi}(v_1) = \operatorname{diag}(u_0, s(p))$. Let z in $M_{2n}(\mathbb{C})$ be the self-adjoint unitary matrix given by

$$z = \begin{pmatrix} 1_n - s(p) & s(p) \\ s(p) & 1_n - s(p) \end{pmatrix},$$

and put $v = z v_1 z^*$. Then

$$\widetilde{\psi}(v) = z\widetilde{\psi}(v_1)z^* = z \begin{pmatrix} u_0 & 0 \\ 0 & s(p) \end{pmatrix} z^* = \begin{pmatrix} u & 0 \\ 0 & 0_n \end{pmatrix}.$$

It now follows from Proposition 9.2.2 that

$$\delta_1([u]_1) = [1_{2n} - v^*v]_0 - [1_{2n} - vv^*]_0 = [1_{2n} - v_1^*v_1]_0 - [1_{2n} - v_1v_1^*]_0$$

$$= \left[\begin{pmatrix} p & 0 \\ 0 & 1_n - s(p) \end{pmatrix} \right]_0 - \left[\begin{pmatrix} s(p) & 0 \\ 0 & 1_n - s(p) \end{pmatrix} \right]_0$$

$$= [p]_0 - [s(p)]_0 = g$$

in $K_0(I)$, and this proves the lemma. $\square$

Combining Propositions 4.3.2, 8.2.4, and 9.1.4, and Lemmas 9.3.2 and 9.3.1, we get the following result.

Proposition 9.3.3. *For every short exact sequence*

$$0 \longrightarrow I \overset{\varphi}{\longrightarrow} A \overset{\psi}{\longrightarrow} B \longrightarrow 0$$

of C^*-algebras, the index map δ_1 makes the sequence of K-groups

$$
\begin{array}{ccccc}
K_1(I) & \overset{K_1(\varphi)}{\longrightarrow} & K_1(A) & \overset{K_1(\psi)}{\longrightarrow} & K_1(B) \\
& & & & \downarrow{\delta_1} \\
K_0(B) & \underset{K_0(\psi)}{\longleftarrow} & K_0(A) & \underset{K_0(\varphi)}{\longleftarrow} & K_0(I)
\end{array}
$$

exact.

9.4* Fredholm operators and Fredholm index

In this section H is an infinite dimensional separable Hilbert space. Let as usual $\mathcal{K}$ denote the C^*-algebra of all compact operators on H, let $\mathcal{Q}(H) = B(H)/\mathcal{K}$ denote the Calkin algebra of H, and let $\pi \colon B(H) \to \mathcal{Q}(H)$ be the quotient mapping. Then we have a short exact sequence

$$(9.10) \qquad 0 \longrightarrow \mathcal{K} \overset{\iota}{\longrightarrow} B(H) \overset{\pi}{\longrightarrow} \mathcal{Q}(H) \longrightarrow 0.$$

Theorem 9.4.1 (Atkinson). *Let T be an operator in $B(H)$. Then the following are equivalent:*

  (i) *$T(H)$ is closed, $\dim(\operatorname{Ker}(T)) < \infty$, and $\dim(\operatorname{Ker}(T^*)) < \infty$,*

  (ii) *there exists an operator S in $B(H)$ such that $1 - ST$ and $1 - TS$ are compact operators,*

  (iii) *$\pi(T)$ is invertible in $\mathcal{Q}(H)$.*

See for example [31, Theorem 3.3.11] for a proof of Theorem 9.4.1.

A *Fredholm operator* on H is an operator that satisfies one, and hence all, of the equivalent conditions in Theorem 9.4.1. Let $\Phi(H)$ denote the set of all Fredholm operators on H.

Define the *Fredholm index* of a Fredholm operator T by

$$\text{index}(T) = \dim(\text{Ker}(T)) - \dim(\text{Ker}(T^*)) \in \mathbb{Z}.$$

As shown in Corollary 6.4.2, $K_0(\mathcal{K})$ is isomorphic to $\mathbb{Z}$, and there is an isomorphism $K_0(\text{Tr}) \colon K_0(\mathcal{K}) \to \mathbb{Z}$ which satisfies $K_0(\text{Tr})([E]_0) = \text{Tr}(E) = \dim(E(H))$ for all projections E in $\mathcal{K}$.

The proposition below says that the index map generalizes the Fredholm index. The proof uses that two Fredholm operators have the same Fredholm index if they are homotopic inside $\Phi(H)$, see for example [29, Theorem 1.4.17] for this result. Consult Remark 8.1.7 for the definition of the K_1-class of an invertible (possibly non-unitary) element.

Proposition 9.4.2. *For each Fredholm operator T on H,*

$$\text{index}(T) = (K_0(\text{Tr}) \circ \delta_1)([\pi(T)]_1).$$

Proof. Let $T = S|T|$ be the polar decomposition of T, where S is a partial isometry in $B(H)$, and where $|T| = (T^*T)^{1/2}$. Put $A = \pi(T)$ and $U = \pi(S)$. Then A is invertible in $\mathcal{Q}(H)$ by Atkinson's theorem because T is Fredholm. Since $A = U|A|$ is the polar decomposition of A in $\mathcal{Q}(H)$, it follows that U is unitary in $\mathcal{Q}(H)$. Thus $\pi(I - S^*S) = \pi(I - SS^*) = 0$, and $E = I - S^*S$ and $F = I - SS^*$ are projections in $\mathcal{K}$.

For t in $[0,1]$ define $R_t = S(t|T| + (1-t)I)$. Then $\pi(R_t) = U(t|A| + (1-t)I)$ is invertible in $\mathcal{Q}(H)$, so that R_t is a Fredholm operator for each t in $[0,1]$. Hence $t \mapsto R_t$ is a continuous path in $\Phi(H)$ connecting $S = R_0$ to $T = R_1$. Also, $A \sim_h U$ in $\text{GL}(\mathcal{Q}(H))$, because $t \mapsto \pi(R_t)$ is a continuous path in $\text{GL}(\mathcal{Q}(H))$ from A to U. This entails that $\text{index}(T) = \text{index}(S)$ and that $[A]_1 = [U]_1$ by Remark 8.1.7. Since $E(H) = \text{Ker}(S)$ and $F(H) = \text{Ker}(S^*)$, Proposition 9.2.4 yields that

$$
\begin{aligned}
(K_0(\text{Tr}) \circ \delta_1)([\pi(T)]_1) &= (K_0(\text{Tr}) \circ \delta_1)([A]_1) = (K_0(\text{Tr}) \circ \delta_1)([U]_1) \\
&= K_0(\text{Tr})([E]_0 - [F]_0) = \text{Tr}(E) - \text{Tr}(F) \\
&= \dim(E(H)) - \dim(F(H)) = \text{index}(S) = \text{index}(T),
\end{aligned}
$$

as desired   □

Example 9.4.3 (The K_1-group of the Calkin algebra). The short exact sequence (9.10) induces the exact sequence

$$K_1(\mathcal{K}) \longrightarrow K_1(B(H)) \longrightarrow K_1(\mathcal{Q}(H))$$
$$\Big\downarrow \delta_1$$
$$K_0(\mathcal{Q}(H)) \longleftarrow K_0(B(H)) \longleftarrow K_0(\mathcal{K}).$$

From Examples 3.3.3 and 8.1.8 we know that $K_0(B(H)) = 0$ and $K_1(B(H)) = 0$. Thus the index map δ_1 is an isomorphism. It follows that

$$K_1(\mathcal{Q}(H)) \cong K_0(\mathcal{K}) \cong \mathbb{Z}.$$

Example 9.4.4 (The Toeplitz algebra). Let $\{e_n\}_{n=1}^{\infty}$ be an orthonormal basis for our Hilbert space H. Let S be the bounded operator on H given by $Se_n = e_{n+1}$ for each n in $\mathbb{N}$, see Example 2.2.9. Then S is an isometry, i.e., $S^*S = I$, where I denotes the identity operator on H, $S^*e_1 = 0$, and $S^*e_n = e_{n-1}$ for $n \geqslant 2$. The operator S is called the *unilateral shift* (with respect to the orthonormal basis $\{e_n\}_{n=1}^{\infty}$).

Define the *Toeplitz algebra* $\mathcal{T}$ to be $C^*(S)$, that is, the sub-C^*-algebra of $B(H)$ generated by S. The Toeplitz algebra has the following universal property established by L. Coburn in [10]: if A is a unital C^*-algebra and if v is an isometry in A, then there is one and only one *-homomorphism $\varphi : \mathcal{T} \to A$ with $\varphi(S) = v$. Moreover, φ is an isomorphism if and only if v is a non-unitary isometry. The Toeplitz algebra is therefore in this sense the universal C^*-algebra generated by a non-unitary isometry.

For i, j in $\mathbb{N}$, let E_{ij} be the bounded operator on H given by $E_{ij}x = \langle x, e_j \rangle e_i$, where $\langle \cdot, \cdot \rangle$ is the inner product on H, and put

$$F_n = \sum_{j=1}^{n} E_{jj}.$$

Then F_n is the projection onto the subspace H_n of H spanned by $\{e_1, \ldots, e_n\}$, and

$$B(H_n) = F_n B(H) F_n = \mathrm{span}\{E_{ij} : 1 \leqslant i, j \leqslant n\}.$$

Observe that

$$F_1 = E_{11} = I - SS^*, \qquad E_{ij} = S^{i-1} F_1 (S^*)^{j-1}, \qquad i, j \in \mathbb{N}.$$

Hence each E_{ij} belongs to $\mathcal{T}$. The C^*-algebra of all compact operators $\mathcal{K}$ on H is the closure of the union $\bigcup_{n=1}^{\infty} F_n B(H) F_n$ (see [31, Lemma 3.3.2 and Theorem 3.3.3]), and it is therefore contained in $\mathcal{T}$. Consequently, $\mathcal{K}$ is an ideal in $\mathcal{T}$.

Consider the quotient mapping $\pi \colon B(H) \to \mathcal{Q}(H)$. Since $F_1 = I - SS^*$ belongs to $\mathcal{K}$, $\pi(S)$ is a unitary element in $\mathcal{Q}(H)$. We claim that $\mathrm{sp}(\pi(S)) = \mathbb{T}$. Indeed, S is a Fredholm operator (because $\pi(S)$ is invertible), $\mathrm{Ker}(S) = 0$, and $\mathrm{Ker}(S^*) = \mathrm{span}\{e_1\}$. This entails that $\mathrm{index}(S) = -1$. It follows from Proposition 9.4.2 that $[\pi(S)]_1$ is a non-zero element of $K_1(\mathcal{Q}(H))$, and so $\pi(S)$ is not homotopic to the unit in $\mathcal{U}(\mathcal{Q}(H))$. By Lemma 2.1.3 (ii) we conclude that $\mathrm{sp}(\pi(S)) = \mathbb{T}$, as claimed.

Notice that $\mathcal{T}/\mathcal{K} = C^*(\pi(S))$ and that $C^*(\pi(S))$ is isomorphic to $C(\mathbb{T}) = C(\mathrm{sp}(\pi(S)))$. We therefore have the short exact sequence

$$(9.11) \qquad 0 \longrightarrow \mathcal{K} \overset{\iota}{\longrightarrow} \mathcal{T} \overset{\psi}{\longrightarrow} C(\mathbb{T}) \longrightarrow 0,$$

where ψ is the composition of π with the isomorphism from $C^*(\pi(S))$ onto $C(\mathbb{T})$ that maps $\pi(S)$ to the identity function in $C(\mathbb{T})$, so that $\psi(S)(z) = z$ for all z in $\mathbb{T}$.

The K-theory of the Toeplitz algebra will be calculated in Exercise 12.4; we can here reveal that $K_0(\mathcal{T})$ is isomorphic to $\mathbb{Z}$ and that $K_1(\mathcal{T}) = 0$. At this time we shall make the following observation about the index map that will be the key ingredient in the calculation of the K-theory of $\mathcal{T}$. By Proposition 9.2.4 and the fact that $\psi(S)$ is the identity function, we have

$$\delta_1([\psi(S)]_1) = [I - S^*S]_0 - [I - SS^*]_0 = -[F_1]_0.$$

Notice that $-[F_1]_0$ generates the group $K_0(\mathcal{K})$, because $K_0(\mathrm{Tr})(-[F_1]_0) = -1$ is a generator of $\mathbb{Z}$.

The Toeplitz algebra $\mathcal{T}$ is infinite in the sense of Definition 5.1.1 because S is left-invertible, but not invertible.

Consider the *reduced Toeplitz algebra* $\mathcal{T}_0$ given by $\mathcal{T}_0 = \psi^{-1}(C_0(\mathbb{T} \setminus \{1\}))$, where $C_0(\mathbb{T} \setminus \{1\})$ is the set of all functions f in $C(\mathbb{T})$ with $f(1) = 0$. We have a short exact sequence

$$(9.12) \qquad 0 \longrightarrow \mathcal{K} \overset{\iota}{\longrightarrow} \mathcal{T}_0 \overset{\psi}{\longrightarrow} C_0(\mathbb{T} \setminus \{1\}) \longrightarrow 0.$$

The K-theory of the reduced Toeplitz algebra vanishes: $K_0(\mathcal{T}_0) = 0$ and

$K_1(\mathcal{T}_0) = 0$, see Exercise 12.4.

The reduced Toeplitz algebra $\mathcal{T}_0$ is non-unital because it has the non-unital C^*-algebra $C_0(\mathbb{T} \setminus \{1\})$ as a quotient. Since $\mathcal{T}_0 + \mathbb{C}I = \mathcal{T}$, we may identify $\widetilde{\mathcal{T}_0}$ with $\mathcal{T}$. Recall from Definition 5.1.1 that a non-unital C^*-algebra is infinite if (and only if) its unitization is infinite. It follows that $\mathcal{T}_0$ is infinite. However, all projections in $\mathcal{T}_0$ are finite, see Definition 5.1.1.

To see this, notice that if F is a projection in $\mathcal{T}_0$, then $\psi(F)$ is a projection in $C_0(\mathbb{T} \setminus \{1\})$, and hence $\psi(F) = 0$, which shows that F belongs to $\mathcal{K}$. But each projection F in $\mathcal{K}$ is finite, because if $F \sim F_0 \leqslant F$, then $\mathrm{Tr}(F - F_0) = \mathrm{Tr}(F) - \mathrm{Tr}(F_0) = 0$, and this entails that $F - F_0 = 0$.

9.5 Exercises

Exercise 9.1. Prove split exactness of K_0 (Proposition 4.3.3) using Proposition 9.3.3.

Exercise 9.2. Let

$$0 \longrightarrow I \overset{\varphi}{\longrightarrow} A \overset{\psi}{\longrightarrow} B \longrightarrow 0$$

be a short exact sequence, where A and B are unital C^*-algebras. Let u be in $\mathcal{U}_n(B)$. Show that $\delta_1([u]_1) = 0$ if and only if there exist an integer $m \geqslant n$ and a unitary v in $\mathcal{U}_m(A)$ such that $\psi(v) = u \oplus 1_{m-n}$. [Hint: See Exercise 8.6.]

Exercise 9.3. Let A be a unital C^*-algebra.

(i) Let a be an element in A with $\|a\| \leqslant 1$. Notice that $1 - a^*a$ and $1 - aa^*$ are positive elements and justify that $af(a^*a) = f(aa^*)a$ for every continuous function $f \colon \mathbb{R}^+ \to \mathbb{C}$. Use this to show that

$$(9.13) \qquad v = \begin{pmatrix} a & (1 - aa^*)^{1/2} \\ -(1 - a^*a)^{1/2} & a^* \end{pmatrix}$$

is a unitary element in $M_2(A)$.

(ii) Let I be an ideal in A, and let $\pi \colon A \to A/I$ be the quotient mapping. Let u be a unitary element in A/I. By Paragraph 2.2.10 there is an element a in A satisfying $\|a\| = 1$ and $\pi(a) = u$. Define v in $\mathcal{U}_2(A)$ by (9.13). Show that

$$\pi(v) = \begin{pmatrix} u & 0 \\ 0 & u^* \end{pmatrix}.$$

(iii) Let $\delta_1 \colon K_1(A/I) \to K_0(I)$ be the index map associated with the short exact sequence

$$0 \longrightarrow I \xrightarrow{\iota} A \xrightarrow{\pi} A/I \longrightarrow 0.$$

Let u be a unitary element in A/I. Use (ii) to give an explicit expression, in terms of a, for a projection p in $M_2(\tilde{I})$ for which

$$\delta_1([u]_1) = [p]_0 - [s(p)]_0.$$

(iv) As in (ii), let u be a unitary element in A/I and let a be a lift of u with $\|a\| = 1$. Let v be the partial isometry in $M_2(A)$ that lifts $\mathrm{diag}(u,0)$ constructed in the proof of Lemma 9.2.1 based on the lift a of u. Calculate $1_2 - v^*v$ and $1_2 - vv^*$. Compare the result with Proposition 9.2.2 and (iii).

Exercise 9.4. Let H be an infinite dimensional separable Hilbert space.

(i) Let T be a Fredholm operator on H, and let K be a compact operator on H. Use Theorem 9.4.1 to show that $T + K$ is a Fredholm operator, and show next that $\mathrm{index}(T + K) = \mathrm{index}(T)$.

(ii) Show that $\mathrm{index}(T) = 0$ for each normal Fredholm operator T.

(iii) Let S be the unilateral shift (from Example 9.4.4). Show that $S^*S - SS^*$ is a compact operator, and that there exists no normal operator N such that $S - N$ is compact. With $\pi \colon B(H) \to \mathcal{Q}(H)$ the quotient mapping, show that $\pi(S)$ is a normal element in $\mathcal{Q}(H)$ and that $\pi(S)$ has no lift to a normal element in $B(H)$.

Exercise 9.5. Consider the short exact sequence

$$0 \longrightarrow C_0(\mathbb{R}^2) \xrightarrow{\varphi} C(\mathbb{D}) \xrightarrow{\psi} C(\mathbb{T}) \longrightarrow 0,$$

see Exercise 2.12.

(i) Show that $K_1(C(\mathbb{D})) = 0$, that $K_0(C(\mathbb{D})) \cong \mathbb{Z}$, and that $K_0(\psi)$ is injective. Conclude that $\delta_1 \colon K_1(C(\mathbb{T})) \to K_0(C_0(\mathbb{R}^2))$ is an isomorphism and that $K_0(C_0(\mathbb{R}^2))$ is non-zero.

(ii) Recall from Example 3.3.5 that we have a surjective group homomorphism dim: $K_0(C(S^2)) \to \mathbb{Z}$ given by

$$\dim([p]_0 - [q]_0) = \operatorname{Tr}(p(x_0)) - \operatorname{Tr}(q(x_0)).$$

(Here, x_0 is an arbitrary point in S^2, and the definition of dim is independent of the choice of x_0.) Show that dim is not injective by considering the short exact sequence

$$0 \longrightarrow C_0(\mathbb{R}^2) \xrightarrow{\ \iota\ } C(S^2) \xrightarrow{\ \pi\ } C(\{x_0\}) \longrightarrow 0.$$

(iii) Recall the definition of K_{00} from Paragraph 3.1.5. Show that

$$K_{00}(C_0(\mathbb{R}^2)) \xrightarrow{\ K_{00}(\iota)\ } K_{00}(C(S^2)) \xrightarrow{\ K_{00}(\pi)\ } K_{00}(C(\{x_0\}))$$

is not exact at $K_{00}(C(S^2))$. Conclude that the functor K_{00} is not half exact; see Example 3.3.9.

(iv) Identify $C_0(\mathbb{R}^2)$ with the kernel of ψ, and its unitization with the set of f in $C(\mathbb{D})$ for which $f|_{\mathbb{T}}$ is a constant function. Show that if u in $C(\mathbb{T})$ is the unitary element given by $u(z) = z$, then $\delta_1([u]_1) = [e]_0 - [f]_0$, where

$$e(z) = \begin{pmatrix} |z|^2 & z(1 - |z|^2)^{1/2} \\ \bar{z}(1 - |z|^2)^{1/2} & 1 - |z|^2 \end{pmatrix}, \qquad f(z) = \begin{pmatrix} 1 & 0 \\ 0 & 0 \end{pmatrix}.$$

Compare the result with Example 3.3.9 and Exercise 3.8. [Hint: Use Exercise 9.3.]

Exercise 9.6. Let H be an infinite dimensional separable Hilbert space. Consider the short exact sequence

$$0 \longrightarrow \mathcal{K} \xrightarrow{\ \iota\ } B(H) \xrightarrow{\ \pi\ } \mathcal{Q}(H) \longrightarrow 0,$$

and let $\delta_1 \colon K_1(\mathcal{Q}(H)) \to K_0(\mathcal{K})$ be its associated index map.

(i) Let E, F be projections in $B(H)$ such that $\dim(E(H)) \leqslant \dim(F(H))$. Show that there exists a partial isometry V in $B(H)$ satisfying $V^*V = E$ and $VV^* \leqslant F$.

(ii) Let u be a unitary element in $\mathcal{Q}(H)$. Show that there is S in $B(H)$ such that $\pi(S) = u$ and such that S is either an isometry, i.e., $S^*S = I$, or a co-isometry, i.e., $SS^* = I$.

(iii) Let again u be a unitary element in $\mathcal{Q}(H)$. Show that u lifts to a unitary operator in $B(H)$ (i.e., there is a unitary operator V on H such that $\pi(V) = u$) if and only if $\delta_1([u]_1) = 0$.

Exercise 9.7. Let H be an infinite dimensional separable Hilbert space, and consider the short exact sequence

$$0 \longrightarrow \mathcal{K} \stackrel{\iota}{\longrightarrow} B(H) \stackrel{\pi}{\longrightarrow} \mathcal{Q}(H) \longrightarrow 0.$$

Let $\mathcal{E}$ be a sub-C^*-algebra of $B(H)$ which contains the compact operators $\mathcal{K}$ and the unit, I, of $B(H)$.

(i) Explain that $\mathcal{K}$ is an ideal in $\mathcal{E}$ and that we have a short exact sequence

(9.14) $$0 \longrightarrow \mathcal{K} \stackrel{\iota}{\longrightarrow} \mathcal{E} \stackrel{\pi}{\longrightarrow} A \longrightarrow 0,$$

where A is the unital C^*-algebra $\mathcal{E}/\mathcal{K}$. Show also that if T is in $B(H)$, then T belongs to $\mathcal{E}$ if and only if $\pi(T)$ belongs to A.

(ii) Let u be a unitary element in A. Use Exercise 9.6 to show that u lifts to an isometry or a co-isometry S in $\mathcal{E}$.

(iii) Suppose that A is finite in the sense of Definition 5.1.1. Let S in $\mathcal{E}$ be given, and assume that S is an isometry or a co-isometry. Show that $\pi(S)$ is unitary.

(iv) Let $\delta_1\colon K_1(A) \to K_0(\mathcal{K})$ be the index map associated with the short exact sequence (9.14). Consider also the map $\alpha\colon K_1(A) \to \mathbf{Z}$ given by $\alpha = K_0(\mathrm{Tr}) \circ \delta_1$. Explain that if T belongs to $\mathcal{E} \cap \Phi(H)$, then $\mathrm{index}(T) = \alpha([\pi(T)]_1)$. Suppose that A is finite, and let S in $\mathcal{E}$ be an isometry or a co-isometry. Show that S is unitary if and only if $\alpha([\pi(S)]_1) = 0$.

(v) Let n be a natural number. Explain that $M_n(\mathcal{E})$ is (isomorphic to) a sub-C^*-algebra of $B(H^n)$ and that we have a short exact sequence

(9.15) $$0 \longrightarrow \mathcal{K}(H^n) \stackrel{\iota}{\longrightarrow} M_n(\mathcal{E}) \stackrel{\pi}{\longrightarrow} M_n(A) \longrightarrow 0.$$

(vi) Use the questions (i)–(v) to prove the following proposition.

Proposition. *Let n in $\mathbb{N}$ be given. Let the C^*-algebras $\mathcal{E}$ and A and the homomorphism $\alpha\colon K_1(A) \to \mathbb{Z}$ be given as above. Assume that A is stably finite, see Definition 5.1.1. Then $M_n(\mathcal{E})$ is finite if and only if $\alpha([u]_1) = 0$ for every u in $\mathcal{U}_n(A)$.*

(vii) It is a consequence of the work of Brown, Douglas and Fillmore (the so-called BDF-theory, see [8]), that if A is a unital Abelian separable C^*-algebra then for each homomorphism $\alpha\colon K_1(A) \to \mathbb{Z}$ there is a sub-C^*-algebra $\mathcal{E}$ of $B(H)$, such that $\mathcal{E}$ contains $\mathcal{K}$ and the unit I, and there is a short exact sequence

$$0 \longrightarrow \mathcal{K} \overset{\iota}{\longrightarrow} \mathcal{E} \overset{\pi}{\longrightarrow} A \longrightarrow 0,$$

such that $\alpha = K_0(\mathrm{Tr}) \circ \delta_1$, where δ_1 is the index map associated with the short exact sequence.

Use this fact and question (vi) to construct a finite unital C^*-algebra $\mathcal{E}$ such that $M_2(\mathcal{E})$ is infinite. [Hint: Try with $A = C(\mathbb{T}^3)$. You can use the fact, proved in Example 11.3.4, that $K_1(C(\mathbb{T}^3))$ is isomorphic to $\mathbb{Z}^4$. Compare this with Exercise 8.15.]

Chapter 10

The Higher K-Functors

It is shown that $K_1(A)$ is isomorphic to $K_0(SA)$ for every C^*-algebra A where SA is the suspension of A. The higher K-groups are inductively defined as $K_{n+1}(A) = K_n(SA)$ for every integer $n \geqslant 1$ in agreement with the situation when $n = 0$, and it is shown that any short exact sequence of C^*-algebras gives rise to a long exact sequence of K-groups.

10.1   The isomorphism between $K_1(A)$ and $K_0(SA)$

Recall from Example 4.1.5 that the suspension of a C^*-algebra A is

$$SA = \{f \in C([0,1], A) : f(0) = f(1) = 0\} = C_0((0,1), A).$$

To each *-homomorphism $\varphi \colon A \to B$ between C^*-algebras A and B, one associates a *-homomorphism $S\varphi \colon SA \to SB$ by $(S\varphi(f))(t) = \varphi(f(t))$ for t in $[0,1]$. In this way, as one easily checks, S becomes a functor from the category of C^*-algebras into itself, and S maps the zero objects to zero objects.

Lemma 10.1.1. *Let X be a locally compact Hausdorff space and let A be a C^*-algebra. For f in $C_0(X)$ and a in A, let fa denote the element in $C_0(X, A)$ given by $(fa)(x) = f(x)a$. Then the set*

$$\mathrm{span}\{fa : \ f \in C_0(X), \ a \in A\}$$

is dense in $C_0(X, A)$.

Proof.   Let $X^+ = X \cup \{\infty\}$ be the one-point compactification of X. Then

$$C_0(X, A) = \{f \in C(X^+, A) : f(\infty) = 0\}.$$

Let f be an element in $C_0(X, A)$ and let $\varepsilon > 0$ be given. There is an open covering $U_1, U_2, \ldots, U_k$ of X^+ such that $\|f(x) - f(y)\| \leqslant \varepsilon$ whenever x, y belong to U_j for some j, because X^+ is compact. For each j choose a point x_j in U_j such that $x_j = \infty$ if ∞ belongs to U_j. Let $\{h_j\}_{j=1}^k$ be a partition of the unit subordinate to the open cover $\{U_j\}_{j=1}^k$, i.e., h_j belongs to $C(X^+)$, the support of h_j is contained in U_j, $0 \leqslant h_j \leqslant 1$, and $h_1 + \cdots + h_k = 1$. Then

$$\|f(x)h_j(x) - f(x_j)h_j(x)\| \leqslant \varepsilon h_j(x)$$

for all x in X and all j, and hence

$$\left\| f(x) - \sum_{j=1}^{k} f(x_j) h_j(x) \right\| \leqslant \varepsilon$$

for all x in X. Setting $a_j = f(x_j)$, we have $\|f - \sum_{j=1}^{k} h_j a_j\| \leqslant \varepsilon$. Delete those terms $h_j a_j$ from the sum where ∞ belongs to U_j, and hence where $a_j = 0$. In the remaining terms, h_j belongs to $C_0(X)$. $\square$

Proposition 10.1.2. *The functor S is exact.*

Proof. Given a short exact sequence of C^*-algebras

$$0 \longrightarrow I \stackrel{\varphi}{\longrightarrow} A \stackrel{\psi}{\longrightarrow} B \longrightarrow 0,$$

we must show that

$$0 \longrightarrow SI \stackrel{S\varphi}{\longrightarrow} SA \stackrel{S\psi}{\longrightarrow} SB \longrightarrow 0$$

is a short exact sequence, see Paragraph 3.3.8. The only non-trivial part hereof is to show that $S\psi$ is surjective. It follows from Lemma 10.1.1 that the set

$$\mathrm{span}\{fb : b \in B, \ f \in C_0((0,1))\}$$

is a dense subset of SB, and every element in this dense set belongs to the image of $S\psi$ because $S\psi(af) = \psi(a)f$ for every a in A and every f in $C_0((0,1))$. $\square$

Theorem 10.1.3. *The groups $K_1(A)$ and $K_0(SA)$ are isomorphic for every C^*-algebra A.*

Moreover, there is a collection of isomorphisms $\theta_A\colon K_1(A) \to K_0(SA)$, one for each C^*-algebra A, such that for every pair of C^*-algebras A and B and every $*$-homomorphism $\varphi\colon A \to B$, the diagram

(10.1)

$$
\begin{array}{ccc}
K_1(A) & \xrightarrow{\;K_1(\varphi)\;} & K_1(B) \\[2pt]
\Big\downarrow{\scriptstyle \theta_A} & & \Big\downarrow{\scriptstyle \theta_B} \\[2pt]
K_0(SA) & \xrightarrow[K_0(S\varphi)]{} & K_0(SB)
\end{array}
$$

is commutative.

The isomorphism θ_A has the following concrete description. Let u in $\mathcal{U}_n(\widetilde{A})$ with $s(u) = 1_n$ be given. Let v in $C([0,1], \mathcal{U}_{2n}(\widetilde{A}))$ be such that $v(0) = 1_{2n}$, $v(1) = \mathrm{diag}(u, u^*)$, and $s(v(t)) = 1_{2n}$ for every t in $[0,1]$. Put

$$
p = v \begin{pmatrix} 1_n & 0 \\ 0 & 0 \end{pmatrix} v^*.
$$

Then p is a projection in $\mathcal{P}_{2n}(\widetilde{SA})$, $s(p) = \mathrm{diag}(1_n, 0_n)$, and

$$
\theta_A([u]_1) = [p]_0 - [s(p)]_0.
$$

Each element g in $K_1(A)$ is represented by a unitary u in $\mathcal{U}_n(\widetilde{A})$ with $s(u) = 1_n$ for some n. Indeed, find n and w in $\mathcal{U}_n(\widetilde{A})$ such that $g = [w]_1$, and put $u = ws(w)^*$. Then $s(u) = 1_n$ and $g = [u]_1$ because $s(w) \sim_h 1_n$, see Lemma 2.1.3 (ii).

To each u in $\mathcal{U}_n(\widetilde{A})$ with $s(u) = 1_n$ we can find v in $C([0,1], \mathcal{U}_{2n}(\widetilde{A}))$ with $v(0) = 1_{2n}$, $v(1) = \mathrm{diag}(u, u^*)$, and $s(v(t)) = 1_{2n}$ for every t in $[0,1]$. By the Whitehead lemma (lemma 2.1.5) we can find z in $C([0,1], \mathcal{U}_{2n}(\widetilde{A}))$ with $z(0) = 1_{2n}$ and $z(1) = \mathrm{diag}(u, u^*)$. The function v given by $v(t) = s(z(t))^* z(t)$ has the desired properties.

Proof of theorem. We have a short exact sequence

(10.2)
$$
0 \longrightarrow SA \xrightarrow{\;\iota\;} CA \xrightarrow{\;\pi\;} A \longrightarrow 0,
$$

where CA is the cone of A. The cone CA is homotopy equivalent to 0 as shown in Example 4.1.5. In particular $K_0(CA) = K_1(CA) = 0$. It follows that the index map $\delta_1 \colon K_1(A) \to K_0(SA)$ associated with the short exact sequence (10.2) is an isomorphism by Proposition 9.3.3. Put $\theta_A = \delta_1$.

Every *-homomorphism $\varphi \colon A \to B$ induces a commutative diagram

$$
\begin{array}{ccccccccc}
0 & \longrightarrow & SA & \longrightarrow & CA & \longrightarrow & A & \longrightarrow & 0 \\
  &   & \Big\downarrow{\scriptstyle S\varphi} &   & \Big\downarrow{\scriptstyle C\varphi} &   & \Big\downarrow{\scriptstyle \varphi} &   &   \\
0 & \longrightarrow & SB & \longrightarrow & CB & \longrightarrow & B & \longrightarrow & 0
\end{array}
$$

where $(C\varphi(f))(t) = \varphi(f(t))$ for t in $[0,1]$. Applying naturality of the index map (Proposition 9.1.5) to this diagram yields that the diagram in (10.1) is commutative.

We now turn to the explicit description of θ_A. We shall use the following identifications: a function f in $C([0,1], M_{2n}(\widetilde{A}))$ belongs to $M_{2n}(\widetilde{CA})$ if and only if $s(f(t)) = f(0)$ for each t in $[0,1]$, and f in $C([0,1], M_{2n}(\widetilde{A}))$ belongs to $M_{2n}(\widetilde{SA})$ if and only if $s(f(t)) = f(0) = f(1)$ for each t in $[0,1]$. Notice also that with π as in (10.2), $\widetilde{\pi}(f) = f(1)$ for f in $M_{2n}(\widetilde{CA})$.

With these identifications, v belongs to $\mathcal{U}_{2n}(\widetilde{CA})$ and

$$
\widetilde{\pi}(v) = \begin{pmatrix} u & 0 \\ 0 & u^* \end{pmatrix}, \qquad p = v \begin{pmatrix} 1_n & 0 \\ 0 & 0 \end{pmatrix} v^* \in \mathcal{P}_{2n}(\widetilde{SA}).
$$

By the definition of the index map, $\theta_A([u]_1) = \delta_1([u]_1) = [p]_0 - [s(p)]_0$, as desired. $\square$

10.2 The long exact sequence in K-theory

We define below the higher K-functors, K_n, for every integer $n \geqslant 2$. The Bott periodicity theorem, treated in the next chapter, states that $K_n \cong K_{n-2}$ for each $n \geqslant 2$. For this reason we shall refrain from spending much time developing and recording properties of the higher K-groups; the long exact sequence established in this section is not an end in itself, but should be viewed as a step towards the six-term exact sequence obtained in Chapter 12.

Definition 10.2.1. For each integer $n \geqslant 2$ define inductively a functor K_n from the category of C^*-algebras to the category of Abelian groups by $K_n = K_{n-1} \circ S$, where the suspension S is viewed as a functor from the category of C^*-algebras into itself.

More specifically, for each integer $n \geqslant 2$ and for each C^*-algebra A define

$$K_n(A) = K_{n-1}(SA),$$

and for each *-homomorphism $\varphi\colon A \to B$ between C^*-algebras A and B, define

$$K_n(\varphi) = K_{n-1}(S\varphi)\colon K_n(A) \to K_n(B).$$

Proposition 10.2.2. *For each integer $n \geqslant 2$, K_n is a half exact functor from the category of C^*-algebras to the category of Abelian groups.*

Proof. The suspension S is a functor, i.e., (i) and (ii) in Paragraph 3.2.1 hold. Since K_1 is a functor and the composition of two functors is again a functor, we obtain by induction that K_n is a functor for each $n \geqslant 2$. Half exactness of K_n follows from half exactness of K_{n-1} combined with exactness of S, see Proposition 10.1.2. $\square$

The nth iterated suspension of a C^*-algebra A is denoted by $S^n A$; it is inductively defined by $S^n A = S(S^{n-1}A)$. Similarly, if B is another C^*-algebra and if $\varphi\colon A \to B$ is a *-homomorphism, then we have a *-homomorphism $S^n\varphi\colon S^n A \to S^n B$, inductively defined by $S^n\varphi = S(S^{n-1}\varphi)$. The higher K-groups are given by

$$K_n(A) = K_1(S^{n-1}A) \cong K_0(S^n A),$$

and $K_n(\varphi) = K_1(S^{n-1}\varphi)$ for each integer $n \geqslant 2$. Apply the convention $S^0 A = A$ and $S^0\varphi = \varphi$.

10.2.3 The higher index maps. Let

$$0 \longrightarrow I \xrightarrow{\varphi} A \xrightarrow{\psi} B \longrightarrow 0$$

be a short exact sequence of C^*-algebras. For $n \geqslant 1$ define inductively the index map $\delta_{n+1}\colon K_{n+1}(B) \to K_n(I)$ as follows. By exactness of S (Proposition 10.1.2), and hence of S^n, the sequence

$$(10.3) \qquad 0 \longrightarrow S^n I \xrightarrow{S^n\varphi} S^n A \xrightarrow{S^n\psi} S^n B \longrightarrow 0$$

is exact, and by Theorem 10.1.3 we have an isomorphism

$$\theta_{S^{n-1}I}\colon K_n(I) = K_1(S^{n-1}I) \to K_0(S^n I).$$

Hence there is one and only one group homomorphism δ_{n+1} making the diagram

$$
\begin{array}{ccc}
K_{n+1}(B) & \xrightarrow{\delta_{n+1}} & K_n(I) \\
\| & & \downarrow{\theta_{S^{n-1}I}} \\
K_1(S^n B) & \xrightarrow[\overline{\delta}_1]{} & K_0(S^n I)
\end{array}
$$

commutative, where $\overline{\delta}_1$ is the index map associated with the short exact sequence (10.3).

The index maps $\delta_1, \delta_2, \ldots$ are *natural* in the following sense. Given a commutative diagram

(10.4)
$$
\begin{array}{ccccccccc}
0 & \longrightarrow & I & \xrightarrow{\varphi} & A & \xrightarrow{\psi} & B & \longrightarrow & 0 \\
& & \downarrow{\gamma} & & \downarrow{\alpha} & & \downarrow{\beta} & & \\
0 & \longrightarrow & I' & \xrightarrow[\varphi']{} & A' & \xrightarrow[\psi']{} & B' & \longrightarrow & 0
\end{array}
$$

with short exact rows of C^*-algebras, and where α, β, γ are *-homomorphisms, the diagram

(10.5)
$$
\begin{array}{ccc}
K_{n+1}(B) & \xrightarrow{\delta_{n+1}} & K_n(I) \\
K_{n+1}(\beta)\downarrow & & \downarrow{K_n(\gamma)} \\
K_{n+1}(B') & \xrightarrow[\delta'_{n+1}]{} & K_n(I')
\end{array}
$$

is commutative. To see this, apply the exact functor S^n to the diagram in (10.4), let $\overline{\delta}_1$ and $\overline{\delta}_1'$ be the index maps of the two resulting short exact sequences, and consider the diagram

(10.6)
$$
\begin{array}{ccccccc}
K_{n+1}(B) & = & K_1(S^n B) & \xrightarrow{\overline{\delta}_1} & K_0(S^n I) & \xrightarrow{\theta^{-1}_{S^{n-1}I}} & K_n(I) \\
K_{n+1}(\beta)\downarrow & & K_1(S^n \beta)\downarrow & & \downarrow{K_0(S^n \gamma)} & & \downarrow{K_n(\gamma)} \\
K_{n+1}(B') & = & K_1(S^n B') & \xrightarrow[\overline{\delta}_1']{} & K_0(S^n I') & \xrightarrow[\theta^{-1}_{S^{n-1}I'}]{} & K_n(I').
\end{array}
$$

The center square of this diagram commutes by naturality of the index map δ_1

(see Proposition 9.1.5), and the right-hand square commutes by naturality of θ (see Theorem 10.1.3). Hence (10.6) is a commutative diagram, and this entails that (10.5) is commutative.

Proposition 10.2.4 (The long exact sequence in K-theory). *Every short exact sequence of C^*-algebras*

$$0 \longrightarrow I \xrightarrow{\varphi} A \xrightarrow{\psi} B \longrightarrow 0$$

induces an exact sequence of K-groups:

$$\cdots \xrightarrow{K_{n+1}(\psi)} K_{n+1}(B) \xrightarrow{\delta_{n+1}} K_n(I) \xrightarrow{K_n(\varphi)} K_n(A) \xrightarrow{K_n(\psi)} K_n(B) \xrightarrow{\delta_n} K_{n-1}(I) \xrightarrow{K_{n-1}(\varphi)} \cdots$$

$$\cdots \xrightarrow{\delta_1} K_0(I) \xrightarrow{K_0(\varphi)} K_0(A) \xrightarrow{K_0(\psi)} K_0(B),$$

where δ_1 is the index map and δ_n, for $n \geqslant 2$, its higher analogues.

Proof. Let $\{\bar{\delta}_n\}_{n=1}^{\infty}$ be the index maps associated with the short exact sequence

$$0 \longrightarrow SI \xrightarrow{S\varphi} SA \xrightarrow{S\psi} SB \longrightarrow 0.$$

It follows directly from the definition of the index maps and Theorem 10.1.3 that the diagrams

$$
\begin{array}{ccccccccccc}
K_2(I) & \longrightarrow & K_2(A) & \longrightarrow & K_2(B) & \xrightarrow{\delta_2} & K_1(I) & \longrightarrow & K_1(A) & \longrightarrow & K_1(B) \\
\| & & \| & & \| & & \theta_I \downarrow & & \theta_A \downarrow & & \theta_B \downarrow \\
K_1(SI) & \longrightarrow & K_1(SA) & \longrightarrow & K_1(SB) & \xrightarrow[\bar{\delta}_1]{} & K_0(SI) & \longrightarrow & K_0(SA) & \longrightarrow & K_0(SB)
\end{array}
$$

and, for $n \geqslant 3$,

$$
\begin{array}{ccccccccccc}
K_n(I) & \longrightarrow & K_n(A) & \longrightarrow & K_n(B) & \xrightarrow{\delta_n} & K_{n-1}(I) & \longrightarrow & K_{n-1}(A) & \longrightarrow & K_{n-1}(B) \\
\| & & \| & & \| & & \| & & \| & & \| \\
K_{n-1}(SI) & \longrightarrow & K_{n-1}(SA) & \longrightarrow & K_{n-1}(SB) & \xrightarrow[\bar{\delta}_{n-1}]{} & K_{n-2}(SI) & \longrightarrow & K_{n-2}(SA) & \longrightarrow & K_{n-2}(SB)
\end{array}
$$

are commutative. The lower row in the first diagram is exact by Proposition 9.3.3, and for both diagrams, exactness of the lower row implies exactness of the upper row. Exactness of the long exact sequence is hence established by induction. $\qquad \square$

Example 10.2.5. The suspension SA of a C^*-algebra A is (isomorphic to) $C_0(\mathbb{R}, A)$. To see this, use that $\mathbb{R}$ is homeomorphic to the open interval $(0, 1)$ and that $SA = C_0((0, 1), A)$. Notice that $C_0(X, C_0(Y))$ is isomorphic to $C_0(X \times Y)$ for any pair of locally compact Hausdorff spaces X and Y. Consequently, $S^n\mathbb{C}$ is isomorphic to $C_0(\mathbb{R}^n)$, and hence

$$K_n(\mathbb{C}) \cong K_0(C_0(\mathbb{R}^n)), \qquad K_{n+1}(\mathbb{C}) \cong K_1(C_0(\mathbb{R}^n))$$

for all integers $n \geqslant 1$.

10.3   Exercises

Exercise 10.1. For every C^*-algebra A put $\mathbb{T}A = C(\mathbb{T}, A)$.

(i) Construct a split exact sequence

$$0 \longrightarrow SA \longrightarrow \mathbb{T}A \rightleftarrows A \longrightarrow 0.$$

(ii) Show that $K_n(\mathbb{T}A)$ is isomorphic to $K_n(A) \oplus K_{n+1}(A)$ for every positive integer n.

(iii) Show that $\mathbb{T}^n\mathbb{C}$ is isomorphic to $C(\mathbb{T}^n)$, and use this and (ii) to express $K_0(C(\mathbb{T}^n))$ and $K_1(C(\mathbb{T}^n))$ in terms of the groups $K_m(\mathbb{C})$. You are welcome to do this for only $n = 1, 2, 3$.

Exercise 10.2. The purpose of this exercise is to establish that the suspension functor S is continuous. More specifically, let

(10.7) $$A_1 \xrightarrow{\varphi_1} A_2 \xrightarrow{\varphi_2} A_3 \longrightarrow \cdots$$

be a sequence of C^*-algebras with inductive limit $(A, \{\mu_n\})$, and let $(B, \{\lambda_n\})$ be the inductive limit of the sequence

$$SA_1 \xrightarrow{S\varphi_1} SA_2 \xrightarrow{S\varphi_2} SA_3 \longrightarrow \cdots.$$

(i) Show that the diagram

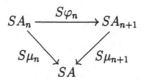

commutes for each positive integer n, and show that there exists a *-homomorphism σ such that the diagram

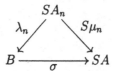

is commutative for each natural number n.

The statement that S is continuous translates into the statement that the *-homomorphism σ above is an isomorphism for every sequence of C^*-algebras (10.7). This is proved in (ii) and (iii) below.

(ii) Show that σ is surjective. [Hint: Use Lemma 10.1.1.]

(iii) Show that $\text{Ker}(S\mu_n) \subseteq \text{Ker}(\lambda_n)$ for each n, and use this to show that σ is injective. [Hint: Show that $\|S\varphi_{m,n}(f)\| \to 0$ as $m \to \infty$ for each f in $\text{Ker}(S\mu_n)$.]

Exercise 10.3. Prove that K_1 is continuous (Proposition 8.2.7). [Hint: Use Theorem 6.3.2, Theorem 10.1.3, and Exercise 10.2.]

Exercise 10.4. Prove that K_1 is stable (Proposition 8.2.8). [Hint: Show first that $K_1(\lambda_{n,A})\colon K_1(A) \to K_1(M_n(A))$ is an isomorphism for each n using Proposition 4.3.8 and Theorem 10.1.3. Then use continuity of K_1 to show that $K_1(\kappa)\colon K_1(A) \to K_1(\mathcal{K}A)$ is an isomorphism.]

Exercise 10.5. Prove that K_1 is split exact (Proposition 8.2.5) using Theorem 10.1.3 and Proposition 4.3.3.

Chapter 11

Bott Periodicity

This chapter contains Atiyah's proof of the analogue in K-theory of the fundamental theorem of R. Bott ([6], see Section 11.4) that the K_0-group of a C^*-algebra A is isomorphic to the K_1-group of the suspension of A. In other words, $K_{n+2}(A)$ is isomorphic to $K_n(A)$ for each integer $n \geqslant 0$. This result is known as Bott periodicity.

11.1 The Bott map

11.1.1 Definition of the Bott map. We shall use the following picture of the suspension SA of a C^*-algebra A:

$$SA = \{f \in C(\mathbb{T}, A) : f(1) = 0\}.$$

Although this is formally different from the definition of the suspension given earlier in this book, the two are clearly isomorphic.

First we consider a unital C^*-algebra A. For every natural number n and every projection p in $\mathcal{P}_n(A)$, define the *projection loop* $f_p \colon \mathbb{T} \to \mathcal{U}_n(A)$ by

$$f_p(z) = zp + (1_n - p) , \qquad z \in \mathbb{T}.$$

By identifying $M_n(\widetilde{SA})$ with the set of functions f in $C(\mathbb{T}, M_n(A))$ such that $f(1)$ belongs to $M_n(\mathbb{C}1_A)$, we obtain that f_p belongs to $\mathcal{U}_n(\widetilde{SA})$. It is easy to check that the following identities hold:

- $f_{p \oplus q} = f_p \oplus f_q$ for all projections p and q in $\mathcal{P}_\infty(A)$,

- $f_0 = 1$,

- if n is a natural number and p and q are projections in $\mathcal{P}_n(A)$ with $p \sim_h q$ in $\mathcal{P}_n(A)$, then $f_p \sim_h f_q$ in $\mathcal{U}_n(\widetilde{SA})$.

Hence we get a group homomorphism $\beta_A \colon K_0(A) \to K_1(SA)$ such that $\beta_A([p]_0) = [f_p]_1$ for every p in $\mathcal{P}_\infty(A)$ from the universal property of K_0 (Proposition 3.1.8). The map β_A is called the *Bott map*.

If $\varphi \colon A \to B$ is a unital *-homomorphism, then

$$\widetilde{S\varphi}(f_p)(z) = \varphi(f_p(z)) = f_{\varphi(p)}(z), \quad z \in \mathbb{T},$$

because $\widetilde{S\varphi}(f)(z) = \varphi(f(z))$ for every f in $M_n(\widetilde{SA})$. This implies that the diagram

(11.1)
$$
\begin{array}{ccc}
K_0(A) & \xrightarrow{K_0(\varphi)} & K_0(B) \\
\beta_A \downarrow & & \downarrow \beta_B \\
K_1(SA) & \xrightarrow[K_1(S\varphi)]{} & K_1(SB)
\end{array}
$$

is commutative. This fact is referred to by saying that the Bott map is *natural*.

Now suppose that A is a non-unital C^*-algebra. Then we have the following diagram with split exact rows:

(11.2)
$$
\begin{array}{ccccccccc}
0 & \longrightarrow & K_0(A) & \longrightarrow & K_0(\widetilde{A}) & \rightleftarrows & K_0(\mathbb{C}) & \longrightarrow & 0 \\
& & \beta_A \downarrow & & \beta_{\widetilde{A}} \downarrow & & \beta_{\mathbb{C}} \downarrow & & \\
0 & \longrightarrow & K_1(SA) & \longrightarrow & K_1(S\widetilde{A}) & \rightleftarrows & K_1(S\mathbb{C}) & \longrightarrow & 0.
\end{array}
$$

The right-hand square is commutative by the commutativity of (11.1). It follows that there is a unique group homomorphism $\beta_A \colon K_0(A) \to K_1(SA)$ making the left-hand square commutative. A straightforward computation yields that

(11.3) $\beta_A([p]_0 - [s(p)]_0) = [f_p f_{s(p)}^*]_1, \quad p \in \mathcal{P}_\infty(\widetilde{A}).$

It follows from (11.3) (or from (11.1) and a diagram chase in (11.2)) that (11.1) holds also in the non-unital case.

Theorem 11.1.2 (Bott periodicity). *The Bott map $\beta_A \colon K_0(A) \to K_1(SA)$ is an isomorphism for every C^*-algebra A.*

In the non-unital case, a diagram chase in (11.2) shows that β_A is an iso-morphism if $\beta_{\widetilde{A}}$ and $\beta_{\mathbb{C}}$ are isomorphisms. Hence, it suffices to prove Theorem 11.1.2 for unital C^*-algebras. This proof is rather long and goes through a number of lemmas. It is given in the following section.

11.2   The proof of Bott periodicity

This section contains a proof of Bott periodicity (Theorem 11.1.2). We have already seen that it suffices to show that β_A is an isomorphism for each *unital* C^*-algebra A. Throughout, A denotes a fixed unital C^*-algebra. For every natural number n consider the following sets:

$$\mathrm{Inv}_0(n) = C(\mathbb{T}, \mathrm{GL}_0(M_n(A))),$$
$$\mathrm{Trig}(n) = \left\{ f \in \mathrm{Inv}_0(n) : f(z) = \textstyle\sum_{j=-m}^{m} a_j z^j,\ m \in \mathbb{N},\ a_j \in M_n(A) \right\},$$
$$\mathrm{Pol}(n,m) = \left\{ f \in \mathrm{Inv}_0(n) : f(z) = \textstyle\sum_{j=0}^{m} a_j z^j,\ a_j \in M_n(A) \right\}, \quad m \in \mathbb{Z}^+,$$
$$\mathrm{Pol}(n) = \textstyle\bigcup_{m=0}^{\infty} \mathrm{Pol}(n,m),$$
$$\mathrm{Lin}(n) = \mathrm{Pol}(n,1),$$
$$\mathrm{Proj}(n) = \{ f_p : p \in \mathcal{P}_n(A) \}.$$

Observe that if $f : \mathbb{T} \to M_n(A)$ is a continuous function such that $f(z)$ is invertible for all z in $\mathbb{T}$ and $f(1) \sim_h 1_n$ in $\mathrm{GL}_n(A)$, then f belongs to $\mathrm{Inv}_0(n)$, since then automatically $f(z)$ belongs to $\mathrm{GL}_0(M_n(A))$ for all z in $\mathbb{T}$.

Note that

$$
\text{(11.4)} \qquad
\begin{aligned}
\mathrm{Proj}(n) &\subseteq\ \mathcal{U}_n(\widetilde{SA}) \ \subseteq\ \mathrm{GL}_n(\widetilde{SA}) \ \subseteq\ \mathrm{Inv}_0(n), \\
\mathrm{Proj}(n) &\subseteq \mathrm{Lin}(n) \subseteq \mathrm{Pol}(n) \subseteq \mathrm{Trig}(n) \subseteq \mathrm{Inv}_0(n).
\end{aligned}
$$

11.2.1 π_0-equivalences. Let Y be a topological space, let X be a subspace of Y, and let $\sim_h$ denote the equivalence relation on X and Y given by homotopy (see Paragraph 2.1.1). The inclusion mapping $\iota : X \to Y$ induces a map $\hat{\iota}$ that makes the diagram

$$
\begin{array}{ccc}
X & \stackrel{\iota}{\hookrightarrow} & Y \\
\downarrow & & \downarrow \\
X/\!\sim_h & \xrightarrow{\ \hat{\iota}\ } & Y/\!\sim_h
\end{array}
$$

commutative. We shall say that ι is a *π_0-equivalence* if $\hat{\iota}$ is bijective. We shall in this Chapter exclusively work with locally path connected spaces, and for such spaces X and Y, a map $\iota\colon X \to Y$ is a π_0-equivalence if and only if it induces a bijection $\pi_0(X) \to \pi_0(Y)$.

The inclusion $\iota\colon X \to Y$ is a π_0-equivalence if and only if

(i) for each y in Y there exists x in X such that $x \sim_h y$ in Y,

(ii) if x_1, x_2 in X are such that $x_1 \sim_h x_2$ in Y, then $x_1 \sim_h x_2$ in X.

If it so happened that the inclusion $\mathrm{Proj}(n) \subseteq \mathcal{U}_n(\widetilde{SA})$ in (11.4) were a π_0-equivalence, then Bott periodicity would follow immediately from this. As it turns out, $\mathrm{Proj}(n) \subseteq \mathcal{U}_n(\widetilde{SA})$ is not in general a π_0-equivalence, but a weaker, and similar, property holds for this inclusion (see Lemma 11.2.13).

We shall study the inclusion $\mathrm{Proj}(n) \subseteq \mathcal{U}_n(\widetilde{SA})$ by studying the six other inclusions in (11.4), and we shall show that each of these inclusions is either a π_0-equivalence or something similar to that.

The inclusion $\mathrm{GL}_n(\widetilde{SA}) \subseteq \mathcal{U}_n(\widetilde{SA})$ is a π_0-equivalence by Proposition 2.1.8, and the lemma below shows that the inclusion $\mathcal{U}_n(\widetilde{SA}) \subseteq \mathrm{Inv}_0(n)$ is also a π_0-equivalence.

Lemma 11.2.2. *Let n be a natural number.*

(i) *For each function f in $\mathrm{Inv}_0(n)$, there is a function g in $\mathrm{GL}_n(\widetilde{SA})$ such that $f \sim_h g$ in $\mathrm{Inv}_0(n)$.*

(ii) *If f and g are functions in $\mathrm{GL}_n(\widetilde{SA})$ with $f \sim_h g$ in $\mathrm{Inv}_0(n)$, then $f \sim_h g$ in $\mathrm{GL}_n(\widetilde{SA})$.*

Proof. (i). We can take a continuous path $t \mapsto a_t$ in $\mathrm{GL}_n(A)$ from $a_0 = 1$ to $a_1 = f(1)$ because $f(1) \sim_h 1$. Set

$$g_t(z) = a_t^{-1} f(z), \qquad z \in \mathbb{T}, \ t \in [0,1].$$

The map $t \mapsto g_t$ in $\mathrm{Inv}_0(n)$ is continuous and $g_1(1) = 1$. It follows that $f = g_0 \sim_h g_1$ in $\mathrm{Inv}_0(n)$ and that g_1 belong to $\mathrm{GL}_n(\widetilde{SA})$.

(ii). Since $f(1)$ and $g(1)$ both belong to the path-wise connected group $\mathrm{GL}_n(\mathbb{C})$, there is a continuous path $t \mapsto a_t$ in $\mathrm{GL}_n(\mathbb{C})$ from $a_0 = f(1)$ to $a_1 = g(1)$. We can by assumption take a continuous path $t \mapsto f_t$ in $\mathrm{Inv}_0(n)$ from $f_0 = f$ to $f_1 = g$. Set

$$g_t(z) = a_t f_t(1)^{-1} f_t(z) \in \mathrm{GL}_n(A), \qquad z \in \mathbb{T}, \ t \in [0,1].$$

Since $g_t(1) = a_t$ and a_t belongs to $M_n(\mathbb{C})$, we see that g_t belongs to $\mathrm{GL}_n(\widetilde{SA})$ for every t in $[0,1]$. Moreover, $t \mapsto g_t$ is continuous, and hence $f = g_0 \sim_h g_1 = g$ in $\mathrm{GL}_n(\widetilde{SA})$. $\square$

Lemma 11.2.3. *The set* $\mathrm{Trig}(n)$ *is dense in* $\mathrm{Inv}_0(n)$ *for each natural number* n.

Proof. Let f in $\mathrm{Inv}_0(n)$ and $\varepsilon > 0$ be given such that $\varepsilon < \|f^{-1}\|_\infty^{-1}$. By Lemma 10.1.1 and its proof there are functions $h_1, h_2, \ldots, h_k$ in $C(\mathbb{T})$ and elements $z_1, z_2, \ldots, z_k$ in $\mathbb{T}$ such that

$$\left\| f(z) - \sum_{j=1}^{k} f(z_j) h_j(z) \right\| \leqslant \varepsilon/2, \qquad z \in \mathbb{T}.$$

By the Stone–Weierstrass theorem the set L consisting of all Laurent polynomials $\sum_{j=-m}^{m} \alpha_j z^j$ is a dense subset of $C(\mathbb{T})$ with respect to the uniform norm $\|\cdot\|_\infty$. Consequently, we can take functions $g_1, \ldots, g_k$ in L such that

$$\|h_j - g_j\|_\infty \leqslant \frac{\varepsilon}{2k\|f(z_j)\|}.$$

Set $g(z) = \sum_{j=1}^{k} f(z_j) g_j(z)$ for z in $\mathbb{T}$. Then $g\colon \mathbb{T} \to M_n(A)$ is continuous and satisfies $\|f - g\|_\infty \leqslant \varepsilon$. Since $\varepsilon < \|f^{-1}\|_\infty^{-1}$, Proposition 2.1.11 shows that g is invertible in $C(\mathbb{T}, M_n(A))$ and that $g(z)$ belongs to $\mathrm{GL}_0(M_n(A))$ for each z in $\mathbb{T}$. It follows that g belongs to $\mathrm{Inv}_0(n)$ and hence to $\mathrm{Trig}(n)$. $\square$

For each natural number k and each function f in $\mathrm{Inv}_0(n)$, denote by $z^k f$ the function given by

$$(z^k f)(z) = z^k f(z), \qquad z \in \mathbb{T}.$$

Clearly $z^k f$ belongs to $\mathrm{Inv}_0(n)$.

Lemma 11.2.4. *Let* n *be a natural number.*

(i) *For every function* f *in* $\mathrm{Inv}_0(n)$, *there are a natural number* k *and a function* g *in* $\mathrm{Pol}(n)$ *such that* $z^k f \sim_h g$ *in* $\mathrm{Inv}_0(n)$.

(ii) *If* f *and* g *are functions in* $\mathrm{Proj}(n)$ *with* $f \sim_h g$ *in* $\mathrm{Inv}_0(n)$, *then there are natural numbers* k *and* m *such that* $z^k f \sim_h z^k g$ *in* $\mathrm{Pol}(n,m)$.

Proof. (i). By Lemma 11.2.3 we can take g_0 in $\mathrm{Trig}(n)$ such that $\|f - g_0\| < \|f^{-1}\|^{-1}$. It follows from Proposition 2.1.11 that $f \sim_h g_0$ in $\mathrm{Inv}_0(n)$. Take k in $\mathbb{N}$ such that $z^k g_0$ belongs to $\mathrm{Pol}(n)$. Then $z^k f \sim_h z^k g_0$ in $\mathrm{Inv}_0(n)$.

(ii). Suppose that $t \mapsto f_t$ is a continuous path in $\mathrm{Inv}_0(n)$ from $f_0 = f$ to $f_1 = g$. Set

$$\delta = \min\{\|f_t^{-1}\|^{-1} : t \in [0,1]\} > 0,$$

and take real numbers $0 = t_0 < t_1 < \cdots < t_N = 1$ such that $\|f_{t_j} - f_{t_{j-1}}\|_\infty < \delta/4$ for each j. Set $g_0 = f_0$ and $g_1 = f_1$. By Lemma 11.2.3 we can find functions $g_{t_1}, \ldots, g_{t_{N-1}}$ in $\mathrm{Trig}(n)$ such that $\|g_{t_j} - f_{t_j}\| \leqslant \delta/4$ for each j. Note that $\|g_{t_{j+1}} - g_{t_j}\| < 3\delta/4$. Proposition 2.1.11 implies that

$$\|g_{t_j}^{-1}\|^{-1} \geqslant \|f_{t_j}^{-1}\|^{-1} - \|g_{t_j} - f_{t_j}\| \geqslant 3\delta/4 > \|g_{t_j} - g_{t_{j-1}}\|$$

for each j. Again by Proposition 2.1.11 (and its proof), if we put

$$g_t = \frac{(t - t_{j-1})g_{t_j} + (t_j - t)g_{t_{j-1}}}{t_j - t_{j-1}}, \qquad t_{j-1} < t < t_j, \ j = 1, 2, \ldots, N,$$

then g_t is invertible in $C(\mathbb{T}, \mathrm{GL}_n(A))$ for each t in $[0,1]$, and $t \mapsto g_t$ is continuous. The element $g_t(z)$ belongs to $\mathrm{GL}_0(M_n(A))$ for each z in $\mathbb{T}$ and each t in $[0,1]$ because the map $s \mapsto g_{st}(z)$ is a continuous path in $\mathrm{GL}_n(A)$ from $g_0(z) = f_0(z)$ in $\mathrm{GL}_0(M_n(A))$ to $g_t(z)$.

Since $g_{t_0}, \ldots, g_{t_N}$ belong to $\mathrm{Trig}(n)$, we can take natural numbers k and m such that $z^k g_{t_j}$ belong to $\mathrm{Pol}(n, m)$ for each j. Then $z^k g_t$ belongs to $\mathrm{Pol}(n, m)$ for each t in $[0,1]$, and therefore

$$z^k f = z^k f_0 = z^k g_0 \sim_h z^k g_1 = z^k f_1 = z^k g$$

in $\mathrm{Pol}(n, m)$. $\square$

The next result has been traced back to G. Higman, see [24, the proof of Theorem 15].

Lemma 11.2.5 (Higman's linearization trick). *For each pair n, m of natural numbers there is a continuous map*

$$\mu_{n,m} \colon \mathrm{Pol}(n, m) \to \mathrm{Lin}((m+1)n)$$

satisfying $f \oplus 1_{mn} \sim_h \mu_{n,m}(f)$ in $\mathrm{Pol}((m+1)n)$ for every function f in $\mathrm{Pol}(n, m)$.

If f belongs to the subset $\mathrm{Proj}(n)$ of $\mathrm{Pol}(n, m)$, then $f \oplus 1_{mn} \sim_h \mu_{n,m}(f)$ in $\mathrm{Lin}((m+1)n)$.

Proof. Suppose that the function f in $\mathrm{Pol}(n, m)$ is given by $f(z) = \sum_{j=0}^{m} a_j z^j$, where $a_0, \ldots, a_m$ belong to $M_n(A)$. Define $\mu_{n,m}$ by

$$\mu_{n,m}(f)(z) = \begin{pmatrix} a_0 & a_1 & a_2 & \cdots & a_{m-1} & a_m \\ -z & 1 & 0 & \cdots & 0 & 0 \\ 0 & -z & 1 & \cdots & 0 & 0 \\ \vdots & \vdots & \vdots & \ddots & \vdots & \vdots \\ 0 & 0 & 0 & \cdots & 1 & 0 \\ 0 & 0 & 0 & \cdots & -z & 1 \end{pmatrix} \in M_{(m+1)n}(A), \quad z \in \mathbf{T}.$$

The continuity of $\mu_{n,m}$ is easily checked using the inequalities (1.4). Set $g_k(z) = \sum_{j=k}^m a_j z^{j-k}$ for $k = 0, 1, \ldots, m$, and set

$$G = \begin{pmatrix} 1 & -g_1 & -g_2 & \cdots & -g_m \\ 0 & 1 & 0 & \cdots & 0 \\ 0 & 0 & 1 & \cdots & 0 \\ \vdots & \vdots & \vdots & \ddots & \vdots \\ 0 & 0 & 0 & \cdots & 1 \end{pmatrix}, \quad H(z) = \begin{pmatrix} 1 & 0 & 0 & \cdots & 0 \\ z & 1 & 0 & \cdots & 0 \\ z^2 & z & 1 & \cdots & 0 \\ \vdots & \vdots & \vdots & \ddots & \vdots \\ z^m & z^{m-1} & z^{m-2} & \cdots & 1 \end{pmatrix}.$$

Note that G and H are invertible (see Exercises 1.14 and 2.3), and consequently G and H belong to $\mathrm{Pol}((m+1)n)$. A matrix multiplication shows that

$$\mu_{n,m}(f)H = \begin{pmatrix} g_0 & g_1 & g_2 & \cdots & g_m \\ 0 & 1 & 0 & \cdots & 0 \\ 0 & 0 & 1 & \cdots & 0 \\ \vdots & \vdots & \vdots & \ddots & \vdots \\ 0 & 0 & 0 & \cdots & 1 \end{pmatrix},$$

and one more matrix multiplication shows that $G\mu_{n,m}(f)H = f \oplus 1_{mn}$. It follows that $\mu_{n,m}(f)$ is invertible, so that $\mu_{n,m}(f)$ belongs to $\mathrm{Inv}_0((m+1)n)$ and hence to $\mathrm{Lin}((m+1)n)$. Moreover, we deduce that $f \oplus 1_{mn} \sim_h \mu_{n,m}(f)$ in $\mathrm{Pol}((m+1)n)$ because $G \sim_h 1$ and $H \sim_h 1$ in $\mathrm{Pol}((m+1)n)$.

To prove the last claim, suppose that $f = f_p$ for some p in $\mathcal{P}_n(A)$. Then

$$\mu_{n,m}(f_p)(z) = \begin{pmatrix} 1-p & p & 0 & \cdots & 0 & 0 \\ -z & 1 & 0 & \cdots & 0 & 0 \\ 0 & -z & 1 & \cdots & 0 & 0 \\ \vdots & \vdots & \vdots & \ddots & \vdots & \vdots \\ 0 & 0 & 0 & \cdots & 1 & 0 \\ 0 & 0 & 0 & \cdots & -z & 1 \end{pmatrix}.$$

For every t in $[0,1]$ set

$$
G_t = \begin{pmatrix} 1 & -tp & 0 & \cdots & 0 \\ 0 & 1 & 0 & \cdots & 0 \\ 0 & 0 & 1 & \cdots & 0 \\ \vdots & \vdots & \vdots & \ddots & \vdots \\ 0 & 0 & 0 & \cdots & 1 \end{pmatrix}, \quad H_t(z) = \begin{pmatrix} 1 & 0 & 0 & \cdots & 0 \\ tz & 1 & 0 & \cdots & 0 \\ tz^2 & tz & 1 & \cdots & 0 \\ \vdots & \vdots & \vdots & \ddots & \vdots \\ tz^m & tz^{m-1} & tz^{m-2} & \cdots & 1 \end{pmatrix}.
$$

Then we see that $G_0 = H_0 = 1$, $G_1 = G$, $H_1 = H$, that G_t and H_t are invertible, and that the maps $t \mapsto G_t$ and $t \mapsto H_t$ are continuous. The matrix product $G_t(z)\mu_{n,m}(f_p)(z)H_t(z)$ is equal to

$$
\begin{pmatrix} (1-p) + (2t - t^2)zp & (1-t)p & 0 & \cdots & 0 & 0 \\ -(1-t)z & 1 & 0 & \cdots & 0 & 0 \\ 0 & -(1-t)z & 1 & \cdots & 0 & 0 \\ \vdots & \vdots & \vdots & \ddots & \vdots & \vdots \\ 0 & 0 & 0 & \cdots & 1 & 0 \\ 0 & 0 & 0 & \cdots & -(1-t)z & 1 \end{pmatrix}
$$

for each t in $[0,1]$. It follows that $G_t\mu_{n,m}(f_p)H_t$ belongs to $\mathrm{Lin}((m+1)n)$ for each t in $[0,1]$, and therefore $f_p \oplus 1_{mn} \sim_h \mu_{n,m}(f_p)$ in $\mathrm{Lin}((m+1)n)$. □

Definition 11.2.6. An element e in a C^*-algebra B is called an *idempotent* if $e^2 = e$. We write $\mathcal{I}(B)$ for the set of all idempotent elements in B.

The following result is analogous to Proposition 2.1.8.

Lemma 11.2.7. *Let B be a C^*-algebra.*

(i) *For every idempotent element e in B, $\rho(e) = ee^*(1_{\tilde{B}} + (e - e^*)(e^* - e))^{-1}$ defines a projection in B.*

(ii) *The map $\rho: \mathcal{I}(B) \to \mathcal{P}(B)$ defined in (i) is continuous, $\rho(p) = p$ for every projection p in B, and $\rho(e) \sim_h e$ in $\mathcal{I}(B)$ for every idempotent e in B.*

(iii) *If p and q are projections in B with $p \sim_h q$ in $\mathcal{I}(B)$, then $p \sim_h q$ in $\mathcal{P}(B)$.*

Proof. (i). Set $h = 1_{\tilde{B}} + (e - e^*)(e^* - e)$. Then clearly h is positive and invertible in $\tilde{B}$, and $\rho(e) = ee^*h^{-1}$ belongs to B because B is an ideal in $\tilde{B}$.

It is straightforward to check that

$$h = 1_{\tilde{B}} + ee^* + e^*e - e - e^*, \quad eh = ee^*e = he, \quad e^*h = e^*ee^* = he^*.$$

This implies that $ee^*h = (ee^*)^2 = hee^*$ and thus $ee^*h^{-1} = h^{-1}ee^*$. We conclude that $\rho(e)$ is self-adjoint and that

$$\rho(e)^2 = ee^*h^{-1}ee^*h^{-1} = (ee^*)^2h^{-2} = ee^*h^{-1} = \rho(e),$$

so that $\rho(e)$ is a projection.

(ii). It is clear from the definition that ρ is continuous (recall that the map $z \mapsto z^{-1}$ for z in $\mathrm{GL}(B)$ is continuous) and that $\rho(p) = p$ for every p in $\mathcal{P}(B)$.

To see that $e \sim_h \rho(e)$ in $\mathcal{I}(B)$, set $w_t = 1_{\tilde{B}} - t(e - \rho(e))$ in $\tilde{B}$ for every t in $[0, 1]$. Observe that $(e - \rho(e))^2 = 0$ because $\rho(e)e = e$ and $e\rho(e) = \rho(e)$. Therefore w_t is invertible with inverse $w_t^{-1} = 1_{\tilde{B}} + t(e - \rho(e))$. It follows that $w_t^{-1}ew_t$ belongs to $\mathcal{I}(B)$, and consequently

$$e = w_0^{-1}ew_0 \sim_h w_1^{-1}ew_1 = (1_{\tilde{B}} + (e - \rho(e)))e(1_{\tilde{B}} - (e - \rho(e))) = \rho(e)$$

in $\mathcal{I}(B)$.

(iii). Suppose that $t \mapsto e_t$ in $\mathcal{I}(B)$ is a continuous path from $e_0 = p$ to $e_1 = q$. Then $t \mapsto \rho(e_t)$ is a continuous path in $\mathcal{P}(B)$ from $\rho(e_0) = p$ to $\rho(e_1) = q$ by (i) and (ii). $\qquad\qquad\qquad\qquad\qquad\qquad\qquad\qquad\qquad\square$

Definition 11.2.8. Set

$$\Pi^+ = \{\alpha \in \mathbb{C} : \mathrm{Re}(\alpha) > 1/2\}, \qquad \Pi^- = \{\alpha \in \mathbb{C} : \mathrm{Re}(\alpha) < 1/2\},$$
$$\Pi = \Pi^+ \cup \Pi^-.$$

An element a in a C^*-algebra B is a *generalized idempotent* if $\mathrm{sp}(a)$ is contained in Π. We denote by $\mathrm{GI}(B)$ the set of generalized idempotents in B, and we set $\mathrm{GI}_n(B) = \mathrm{GI}(M_n(B))$ for every natural number n.

This terminology is justified by the fact that for every idempotent element e in a C^*-algebra B, $\mathrm{sp}(e)$ is contained in $\{0, 1\}$ (see Exercise 11.4), and hence in Π.

Our next lemma relies on the holomorphic function calculus which is outlined in Section 11.5, where the reader can also find an explanation of the notation used in the proof of the lemma.

Lemma 11.2.9. *Let B be a unital C^*-algebra, and define a holomorphic function $h \colon \Pi \to \mathbb{C}$ by*

$$h(z) = \begin{cases} 0, & z \in \Pi^-, \\ 1, & z \in \Pi^+. \end{cases}$$

Then

(i) $h(a)$ *is an idempotent for each generalized idempotent a in B,*

(ii) $h(e) = e$ *for each idempotent e in B,*

(iii) $h(a) \sim_h a$ *in* GI(B) *for each generalized idempotent a in B,*

(iv) *if e and f are idempotents in B with $e \sim_h f$ in* GI(B), *then $e \sim_h f$ in $\mathcal{I}(B)$.*

Proof. (i). This is clear because the function h is idempotent, and the map $h \mapsto h(a)$ is an algebra homomorphism.

(ii). Let e be an idempotent element in B. Let γ_j be the positively oriented circle with center j and radius $1/4$ for $j = 0, 1$. Then $\Gamma = \{\gamma_0, \gamma_1\}$ is a contour surrounding sp(e) in Π for every e in $\mathcal{I}(B)$. The definition of h implies that

$$\frac{1}{2\pi i} \int_{\gamma_0} h(z)(z 1_B - e)^{-1}\, dz = 0$$

and

$$\frac{1}{2\pi i} \int_{\gamma_1} h(z)(z 1_B - e)^{-1}\, dz = \mathrm{Ind}_{\gamma_1}(1)e + \mathrm{Ind}_{\gamma_1}(0)(1_B - e) = e,$$

where we have used the fact that

$$(z 1_B - e)^{-1} = (z - 1)^{-1}e + z^{-1}(1_B - e).$$

We conclude that

$$h(e) = \frac{1}{2\pi i}\left(\int_{\gamma_0} h(z)(z 1_B - e)^{-1}\, dz + \int_{\gamma_1} h(z)(z 1_B - e)^{-1}\, dz \right) = e.$$

(iii). Let a in GI(B) be given. For each t in $[0,1]$ define a holomorphic function $h_t \colon \Pi \to \mathbb{C}$ by

$$h_t(z) = \begin{cases} tz, & z \in \Pi^-, \\ t(z - 1) + 1, & z \in \Pi^+, \end{cases}$$

and set $a_t = h_t(a)$. Observe that $h_0 = h$ and $h_1(z) = z$ for all z in Π.
The element a_t belongs to $\mathrm{GI}(B)$ by Theorem 11.5.3 (iii), because $\mathrm{sp}(a_t)$
is contained in $h_t(\Pi)$ which again is contained in Π. The map $t \mapsto a_t$ is
continuous on $[0, 1]$. Indeed, for each t_0 in $[0, 1]$ and each sequence $\{t_n\}_{n=1}^{\infty}$ in
$[0, 1]$ converging to t_0, $\{h_{t_n}\}_{n=1}^{\infty}$ converges uniformly on every compact subset
of Π to h_{t_0}, and therefore Theorem 11.5.3 (iv) implies that a_{t_n} converges to
a_{t_0}. We conclude that $h(a) = a_0 \sim_h a_1 = a$ in $\mathrm{GI}(B)$.

(iv). Suppose that $t \mapsto a_t$ in $\mathrm{GI}(B)$ is a continuous path from $a_0 = e$ to
$a_1 = f$. We claim that there is a compact subset K of Π such that

$$(11.5) \qquad\qquad \bigcup_{t \in [0,1]} \mathrm{sp}(a_t) \subseteq K.$$

To verify this, set

$$R = \max\{\|a_t\| : t \in [0,1]\},$$
$$r = \min\{\|(a_t - (\tfrac{1}{2} + is) \cdot 1_A)^{-1}\|^{-1} : t \in [0,1],\ s \in [-R, R]\},$$

and let K be the set of those complex numbers α for which $|\alpha| \leqslant R$ and
$|\mathrm{Re}(\alpha) - 1/2| \geqslant r$. Then K is a compact set contained in Π. To prove (11.5),
take a complex number α outside K and take t in $[0, 1]$. If $|\alpha| > R$, then
$|\alpha| > \|a_t\|$ and so α does not belong to $\mathrm{sp}(a_t)$. Otherwise $|\alpha| \leqslant R$, in which
case it follows that $\alpha = 1/2 + \delta + is$ for some real numbers δ and s with
$|\delta| < r$ and $|s| \leqslant R$. Consequently,

$$\|(a_t - \alpha 1_B) - (a_t - (\tfrac{1}{2} + is) \cdot 1_A)\| = |\delta| < r \leqslant \|(a_t - (\tfrac{1}{2} + is) \cdot 1_A)^{-1}\|^{-1}.$$

Proposition 2.1.11 shows that $a_t - \alpha 1_B$ is invertible, so that α does not belong
to $\mathrm{sp}(a_t)$. This proves (11.5).

Choose a contour Γ in Π surrounding K and hence $\mathrm{sp}(a_t)$ for every t in
$[0, 1]$, and define

$$e_t = h(a_t) = \frac{1}{2\pi i} \int_\Gamma h(z)(z 1_B - a_t)^{-1}\, dz \in \mathcal{I}(B), \qquad t \in [0,1].$$

The important fact to notice is that we can use the same contour Γ in the
definition of every e_t. An easy computation yields that

$$(z 1_B - a_t)^{-1} - (z 1_B - a_s)^{-1} = (z 1_B - a_t)^{-1}(a_t - a_s)(z 1_B - a_s)^{-1}$$

for all s, t in $[0, 1]$, and hence

$$e_t - e_s = \frac{1}{2\pi i} \int_\Gamma h(z)(z1_B - a_t)^{-1}(a_t - a_s)(z1_B - a_s)^{-1} \, dz.$$

An obvious norm estimate now shows that

$$\|e_t - e_s\|$$
$$\leqslant \frac{1}{2\pi} \int_\Gamma |h(z)| \, \|(z1_B - a_t)^{-1}\| \, \|a_t - a_s\| \, \|(z1_B - a_s)^{-1}\| \, d|z|$$
$$\leqslant \frac{1}{2\pi} \left(\max\{\|(z1_B - a_r)^{-1}\| : z \in \mathrm{Im}\Gamma, \ r \in [0,1]\}\right)^2 \left(\int_\Gamma d|z|\right) \|a_t - a_s\|.$$

It follows that the map $t \mapsto e_t$ is continuous. Using (ii), we conclude that $e = h(e_0) \sim_h h(e_1) = f$ in $\mathcal{I}(B)$. $\square$

Combining Lemmas 11.2.7 and 11.2.9 we obtain that the inclusion $\mathrm{GI}_n(A) \subseteq \mathcal{P}_n(A)$ is a π_0-equivalence:

Corollary 11.2.10. *Let n be a natural number.*

(i) *For every element a in $\mathrm{GI}_n(A)$ there is a projection p in $\mathcal{P}_n(A)$ such that $a \sim_h p$ in $\mathrm{GI}_n(A)$.*

(ii) *If p and q are projections in $\mathcal{P}_n(A)$ with $p \sim_h q$ in $\mathrm{GI}_n(A)$, then $p \sim_h q$ in $\mathcal{P}_n(A)$.*

Define for each a in $M_n(A)$ a function $f_a \colon \mathbb{T} \to M_n(A)$ by $f_a(z) = az + (1_n - a)$. This definition agrees with the previous definition of the projection loop f_p for p in $\mathcal{P}_n(A)$, see Paragraph 11.1.1. Recall that f_a belongs to $\mathrm{Lin}(n)$ if and only if $f_a(z)$ is invertible for all z in $\mathbb{T}$ and $f(1) \sim_h 1_n$ in $\mathrm{GL}_n(A)$.

Lemma 11.2.11. *For every a in $M_n(A)$, the following assertions are equivalent:*

(i) *$f_a \in \mathrm{Lin}(n)$,*

(ii) *$f_a(z) \in \mathrm{GL}_n(A)$ for every z in $\mathbb{T} \setminus \{1\}$,*

(iii) *$a \in \mathrm{GI}_n(A)$.*

Proof. It is trivial that (i) implies (ii), and (ii) implies (i) because $f_a(1) = 1_n$.

To see that (ii) and (iii) are equivalent, observe that

$$f_a(z) = (z-1)a + 1_n = (z-1)(a - (1-z)^{-1}1_n), \qquad z \in \mathbb{T} \setminus \{1\},$$

where 1_n denotes the unit of $M_n(A)$, and

$$\{(1-z)^{-1} : z \in \mathbb{T} \setminus \{1\}\} = \{\alpha \in \mathbb{C} : \mathrm{Re}(\alpha) = \tfrac{1}{2}\}.$$

Consequently, (ii) holds if and only if $a - \alpha 1_n$ is invertible for every complex number α with $\mathrm{Re}(\alpha) = 1/2$. This is clearly equivalent to (iii). $\square$

If f belongs to $\mathrm{Lin}(n)$, then $f = f_a$ for some a in $\mathrm{GI}_n(A)$ if and only if $f(1) = 1_n$ (with 1_n denoting the unit of $M_n(A)$), and in this case, $a = (1_n - f(-1))/2$. If a, b belong to $\mathrm{GI}_n(A)$, then $a \sim_h b$ in $\mathrm{GI}_n(A)$ if and only if $f_a \sim_h f_b$ in $\mathrm{Lin}(n)$. Indeed, to see the "if" part, let $t \mapsto g_t$ be a continuous path in $\mathrm{Lin}(n)$ such that $g_0 = f_a$ and $g_1 = f_b$. Put $h_t(z) = g_t(1)^{-1}g_t(z)$, and put $a_t = (1_n - h_t(-1))/2$. Then $t \mapsto h_t$ is a continuous path in $\mathrm{Lin}(n)$ from f_a to f_b, $h_t(1) = 1_n$ for each t, and $t \mapsto a_t$ is a continuous path in $\mathrm{GI}_n(A)$ from a to b.

The following lemma says that the inclusion $\mathrm{Proj}(n) \subseteq \mathrm{Lin}(n)$ is a π_0-equivalence.

Lemma 11.2.12. *Let n be a natural number.*

(i) *For every f in $\mathrm{Lin}(n)$ there is a projection p in $\mathcal{P}_n(A)$ such that $f \sim_h f_p$ in $\mathrm{Lin}(n)$.*

(ii) *If p and q are projections in $\mathcal{P}_n(A)$ with $f_p \sim_h f_q$ in $\mathrm{Lin}(n)$, then $p \sim_h q$ in $\mathcal{P}_n(A)$.*

Proof. (i). Let f in $\mathrm{Lin}(n)$ be given. Then $f(1)$ is invertible in $M_n(A)$ and homotopic to 1_n in $\mathrm{GL}_n(A)$. Put $g(z) = f(1)^{-1}f(z)$. Then g belongs to $\mathrm{Lin}(n)$ and $f \sim_h g$ in $\mathrm{Lin}(n)$. Moreover, $g(1) = 1_n$, and so $g = f_a$ for some a in $\mathrm{GI}_n(A)$. By Corollary 11.2.10 (i) there is a projection p in $\mathcal{P}_n(A)$ such that $a \sim_h p$ in $\mathrm{GI}_n(A)$. By the argument above, $f \sim_h g = f_a \sim_h f_p$ in $\mathrm{Lin}(n)$.

(ii). If $f_p \sim_h f_q$ in $\mathrm{Lin}(n)$, then $p \sim_h q$ in $\mathrm{GI}_n(A)$ by the argument above, and hence $p \sim_h q$ in $\mathcal{P}_n(A)$ by Corollary 11.2.10 (ii). $\square$

We summarize the results obtained so far in the following lemma.

Lemma 11.2.13. *Let n be a natural number.*

(i) *For every unitary u in $\mathcal{U}_n(\widetilde{SA})$, there are natural numbers $m \geqslant n$ and k and a projection p in $\mathcal{P}_m(A)$ such that $(z^k u) \oplus 1_{m-n} \sim_h f_p$ in $\mathcal{U}_m(\widetilde{SA})$.*

(ii) *If p and q are projections in $\mathcal{P}_n(A)$ with $f_p \sim_h f_q$ in $\mathcal{U}_n(\widetilde{SA})$, then there exist a natural number $m \geqslant n$ and a projection r in $\mathcal{P}_{m-n}(A)$ such that $p \oplus r \sim_h q \oplus r$ in $\mathcal{P}_m(A)$.*

Proof. (i). By Lemma 11.2.4 (i) we can find natural numbers k and m_1 and a function f in $\text{Pol}(n, m_1)$ such that $z^k u \sim_h f$ in $\text{Inv}_0(n)$. Set $m = (m_1 + 1)n$. There is a function g in $\text{Lin}(m)$ with $f \oplus 1_{m-n} \sim_h g$ in $\text{Pol}(m)$ by Lemma 11.2.5, and we can take a projection p in $\mathcal{P}_m(A)$ with $g \sim_h f_p$ in $\text{Lin}(m)$ by Lemma 11.2.12 (i). We conclude that

$$(z^k u) \oplus 1_{m-n} \sim_h f \oplus 1_{m-n} \sim_h g \sim_h f_p \text{ in } \text{Inv}_0(m).$$

Lemma 11.2.2 (ii) and Proposition 2.1.8 (ii) show that this homotopy can be realized in $\mathcal{U}_m(\widetilde{SA})$.

(ii). Suppose that f_p and f_q are homotopic in $\mathcal{U}_n(\widetilde{SA})$ and hence in $\text{Inv}_0(n)$. We can take natural numbers k and m_1 such that $z^k f_p \sim_h z^k f_q$ in $\text{Pol}(n, m_1)$ by Lemma 11.2.4 (ii). Set $m_2 = (k + 1)n$. Since

$$f_{p \oplus 1_{kn}} = f_p \oplus z 1_{kn}, \qquad f_{q \oplus 1_{kn}} = f_q \oplus z 1_{kn}, \qquad z^k 1_n \oplus 1_{kn} \sim_h 1_n \oplus z 1_{kn}$$

in $\text{Pol}(m_2, k)$, we conclude that

$$f_{p \oplus 1_{kn}} = (1_n \oplus z 1_{kn})(f_p \oplus 1_{kn}) \sim_h (z^k 1_n \oplus 1_{kn})(f_p \oplus 1_{kn})$$
$$\sim_h (z^k 1_n \oplus 1_{kn})(f_q \oplus 1_{kn}) \sim_h f_{q \oplus 1_{kn}}$$

in $\text{Pol}(m_2, m_1)$. Set $m = (m_1 + 1)m_2$, and set

$$r = \begin{pmatrix} 1_{kn} & 0 \\ 0 & 0 \end{pmatrix} \in \mathcal{P}_{kn + m_1 m_2}(A).$$

Lemma 11.2.5 now yields that we have the following homotopy in $\text{Lin}(m)$:

$$f_{p \oplus r} = f_{p \oplus 1_{kn}} \oplus 1_{m_1 m_2} \sim_h \mu_{m_2, m_1}(f_{p \oplus 1_{kn}})$$
$$\sim_h \mu_{m_2, m_1}(f_{q \oplus 1_{kn}}) \sim_h f_{q \oplus 1_{kn}} \oplus 1_{m_1 m_2} = f_{q \oplus r}.$$

Part (ii) follows from this and from Lemma 11.2.12 (ii). $\square$

Proof of Theorem 11.1.2. We show that the Bott map β_A is (i) surjective and (ii) injective.

(i). For a given g in $K_1(SA)$ take n in $\mathbb{N}$ and u in $\mathcal{U}_n(\widetilde{SA})$ such that $g = [u]_1$. By Lemma 11.2.13 (i) we can find natural numbers $m \geqslant n$ and k, and a projection p in $\mathcal{P}_m(A)$ such that $(z^k u) \oplus 1_{m-n} \sim_h f_p$ in $\mathcal{U}_m(\widetilde{SA})$. By the Whitehead lemma (Lemma 2.1.5), $f_{1_{nk}} = z 1_{nk} \sim_h z^k 1_n \oplus 1_{nk-n}$ in $\mathcal{U}_{nk}(\widetilde{SA})$, and so

$$\beta_A([p]_0 - [1_{nk}]_0) = [f_p]_1 - [f_{1_{nk}}]_1 = [z^k u]_1 - [z^k 1_n]_1$$
$$= [u]_1 + [z^k 1_n]_1 - [z^k 1_n]_1 = g.$$

(ii). Let g in $K_0(A)$ with $\beta_A(g) = 0$ be given. Find n in $\mathbb{N}$ and p, q in $\mathcal{P}_n(A)$ such that $g = [p]_0 - [q]_0$. Then $[f_p]_1 = [f_q]_1$, which implies that $f_p \oplus 1_{m-n} \sim_h f_q \oplus 1_{m-n}$ in $\mathcal{U}_m(\widetilde{SA})$ for some natural number $m \geqslant n$. Set

$$p_1 = \begin{pmatrix} p & 0 \\ 0 & 0 \end{pmatrix} \in \mathcal{P}_m(A), \qquad q_1 = \begin{pmatrix} q & 0 \\ 0 & 0 \end{pmatrix} \in \mathcal{P}_m(A).$$

Then $f_{p_1} = f_p \oplus 1_{m-n}$ and $f_{q_1} = f_q \oplus 1_{m-n}$, and consequently $f_{p_1} \sim_h f_{q_1}$ in $\mathcal{U}_m(\widetilde{SA})$. It follows from Lemma 11.2.13 (ii) that $p_1 \oplus r \sim_h q_1 \oplus r$ in $\mathcal{P}_k(A)$ for some natural number $k \geqslant m$ and some r in $\mathcal{P}_{k-m}(A)$. We conclude that

$$g = [p]_0 - [q]_0 = [p_1 \oplus r]_0 - [q_1 \oplus r]_0 = 0,$$

and so β_A is injective. $\qquad\qquad\square$

11.3 Applications of Bott periodicity

We shall in this section and in Chapter 12 study some of the applications of Bott periodicity. First of all, Bott periodicity makes it possible to calculate the K-groups of many concrete C^*-algebras.

Corollary 11.3.1. *For every C^*-algebra A and every integer $n \geqslant 0$,*

$$K_{n+2}(A) \cong K_n(A).$$

Proof. In the case where $n = 0$, this is exactly the content of Theorem 11.1.2. The general statement follows by induction on n because

$$K_{n+2}(A) = K_{n+1}(SA) \cong K_{n-1}(SA) = K_n(A).$$

for every $n \geqslant 1$. $\qquad\qquad\square$

Example 11.3.2. We deduce from Corollary 11.3.1 and Example 10.2.5 that for every natural number n,

$$(11.6) \qquad K_0\left(C_0(\mathbb{R}^n)\right) \cong K_n(\mathbb{C}) \cong \begin{cases} K_0(\mathbb{C}) \cong \mathbb{Z}, & n \text{ even}, \\ K_1(\mathbb{C}) = \{0\}, & n \text{ odd}. \end{cases}$$

Similarly we have

$$(11.7) \qquad K_1\left(C_0(\mathbb{R}^n)\right) \cong \begin{cases} \{0\}, & n \text{ even}, \\ \mathbb{Z}, & n \text{ odd}. \end{cases}$$

Example 11.3.3. For each integer $n \geqslant 0$ consider the n-sphere

$$S^n = \{(x_1, x_2, \ldots, x_{n+1}) \in \mathbb{R}^{n+1} : x_1^2 + x_2^2 + \cdots + x_{n+1}^2 = 1\}.$$

The one-point compactification of $\mathbb{R}^n$ is homeomorphic to S^n (for $n \geqslant 1$), and so we have an isomorphism $C_0(\mathbb{R}^n)^\sim \cong C(S^n)$. It follows from Example 11.3.2, Example 4.3.5, and (8.4) that

$$(11.8) \quad K_0(C(S^n)) \cong \begin{cases} \mathbb{Z} \oplus \mathbb{Z}, & n \text{ even}, \\ \mathbb{Z}, & n \text{ odd}, \end{cases} \qquad K_1(C(S^n)) \cong \begin{cases} \{0\}, & n \text{ even}, \\ \mathbb{Z}, & n \text{ odd}. \end{cases}$$

Example 11.3.4. Recall from Exercise 10.1 that $\mathbb{T}A = C(\mathbb{T}, A)$ when A is a C^*-algebra. Exercise 10.1 and Corollary 11.3.1 imply that

$$K_0(\mathbb{T}A) \cong K_1(\mathbb{T}A) \cong K_0(A) \oplus K_1(A)$$

for every C^*-algebra A. Since $\mathbb{T}^n\mathbb{C} \cong C(\mathbb{T}^n)$ by Exercise 10.1 (iii), we see that

$$(11.9) \qquad K_0(C(\mathbb{T}^n)) \cong K_1(C(\mathbb{T}^n)) \cong \mathbb{Z}^{2^{n-1}}.$$

It follows in particular that $K_1(C(\mathbb{T}^3))$ is isomorphic to $\mathbb{Z}^4$. Exercise 8.15 shows that $\mathcal{U}(C(\mathbb{T}^3))/\mathcal{U}_0(C(\mathbb{T}^3))$ is isomorphic to $\mathbb{Z}^3$. This proves that the map

$$\Delta \colon K_1(C(\mathbb{T}^3)) \to \mathcal{U}(C(\mathbb{T}^3))/\mathcal{U}_0(C(\mathbb{T}^3))$$

from Section 8.3 is *not* injective and that the map

$$[\,\cdot\,]_1 \colon \mathcal{U}(C(\mathbb{T}^3)) \to K_1(C(\mathbb{T}^3))$$

is *not* surjective.

This example can also be used to complete Example 2.2.9, where we claimed that unitarily equivalent projections are not necessarily homotopic. The above reasoning shows that there is an element g in $K_1(C(\mathbb{T}^3))$ which is not represented by any unitary element in $C(\mathbb{T}^3)$. Let k be the smallest natural number such that there is a unitary u in $\mathcal{U}_{2k}(C(\mathbb{T}^3))$ with $g = [u]_1$, and set $A = M_k(C(\mathbb{T}^3))$. Then there is no unitary v in $\mathcal{U}(A)$ such that u is homotopic to $\mathrm{diag}(v, 1)$. (A more refined argument shows that we can take k above to be 1.)

Finally observe that $K_1(C(\mathbb{T}))$ is isomorphic to $\mathbb{Z}$, so that the map

$$\Delta \colon K_1(C(\mathbb{T})) \to \mathcal{U}(C(\mathbb{T}))/\mathcal{U}_0(C(\mathbb{T}))$$

is an isomorphism, see Example 8.3.2. Let v be the unitary element in $C(\mathbb{T})$ given by $v(z) = z$. The winding number of v is one. Example 8.3.2 gives an isomorphism $\mathcal{U}(C(\mathbb{T}))/\mathcal{U}_0(C(\mathbb{T})) \to \mathbb{Z}$ given by winding number. We conclude from these findings that $[v]_1$ is a generator for the cyclic group $K_1(C(\mathbb{T}))$.

11.4* Homotopy groups and K-theory

Let A be a unital C^*-algebra. We have previously defined $\mathcal{U}_\infty(A)$ to be the disjoint union of the groups $\mathcal{U}_n(A)$. In this section we (re)define $\mathcal{U}_\infty(A)$ to be the inductive limit of the sequence of groups

$$\mathcal{U}(A) \longrightarrow \mathcal{U}_2(A) \longrightarrow \mathcal{U}_3(A) \longrightarrow \cdots,$$

where the connecting maps $\mathcal{U}_n(A) \to \mathcal{U}_{n+1}(A)$ are given by $u \mapsto \mathrm{diag}(u, 1)$. Viewing $\mathcal{U}_\infty(A)$ in this way, it becomes a group, and each inductive limit map $\lambda_n \colon \mathcal{U}_n(A) \to \mathcal{U}_\infty(A)$ is a group homomorphism. Equip $\mathcal{U}_\infty(A)$ with the metric given by $d(\lambda_n(u), \lambda_n(v)) = \|u - v\|$, where n is chosen such that u, v are in $\mathcal{U}_n(A)$. This is well-defined because the connecting maps $\mathcal{U}_n(A) \to \mathcal{U}_{n+1}(A)$ are isometries.

Let G be a topological group. Define for each $n \geqslant 1$ its nth *homotopy group*, $\pi_n(G)$, to be the set of homotopy equivalence classes of continuous maps $S^n \to G$ mapping a selected base point of S^n to e, the neutral element of G. Let $\pi_0(G)$ be the set of homotopy equivalence classes in G (or, equivalently, the set of homotopy classes of continuous maps $S^0 \to G$ mapping a selected base point of S^0 to e, where we recall that S^0 is the space consisting of two points, one of which is the base point). Define a group operation on $\pi_n(G)$ by $[f] \cdot [g] = [f \cdot g]$, whenever $f, g: S^n \to G$ are continuous functions mapping the base point of S^n to e, and where $(f \cdot g)(x) = f(x)g(x)$ (using the group operation on G).

For $n \geqslant 1$ there is another definition of this group operation on $\pi_n(G)$ that does not use the group structure of G (this operation consists roughly in "pasting together" two continuous base point preserving functions $f, g: S^n \to G$ to obtain a new continuous base point preserving function $f \star g: S^n \to G$). It is known (from its second picture) that $\pi_n(G)$ is an Abelian group when $n \geqslant 2$.

Now, let A be any C^*-algebra, unital or not, and put

$$\mathcal{V}(A) = \{u \in \mathcal{U}_\infty(\widetilde{A}) : s(u) = 1\},$$

where s is the scalar mapping $\mathcal{U}_\infty(\widetilde{A}) \to \mathcal{U}_\infty(\mathbb{C})$. Observe that $\mathcal{V}(A)$ is a topological group. If A is unital, then $\mathcal{V}(A)$ is isomorphic, as a topological group, to $\mathcal{U}_\infty(A)$.

Proposition 11.4.1. *Let A be a C^*-algebra. Then $K_n(A)$ is isomorphic to $\pi_{n-1}(\mathcal{V}(A))$ for each natural number n. In particular,*

$$(11.10) \qquad \pi_n(\mathcal{V}(A)) = \begin{cases} K_0(A), & n \text{ odd,} \\ K_1(A), & n \text{ even.} \end{cases}$$

R. Bott proved in [6, Corollary to Theorem II] that $\pi_{n+2}(\mathcal{U}_\infty(\mathbb{C})) \cong \pi_n(\mathcal{U}_\infty(\mathbb{C}))$, and hence, in effect, that $K_{n+3}(\mathbb{C}) \cong K_{n+1}(\mathbb{C})$ for all $n \geqslant 0$.

Proof of proposition.   The two definitions of $\mathcal{U}_\infty(\widetilde{A})$ (the one given in Chapter 8 and the other given above) coincide in the following sense: for every pair u in $\mathcal{U}_n(\widetilde{A})$ and v in $\mathcal{U}_m(\widetilde{A})$, $u \sim_1 v$ if and only if $\lambda_n(u) \sim_h \lambda_m(v)$ in $\mathcal{U}_\infty(\widetilde{A})$.

Indeed, if $\lambda_n(u) \sim_h \lambda_m(v)$ in $\mathcal{U}_\infty(\widetilde{A})$, then there exist unitaries

$$u_0 = \lambda_n(u), u_1, u_2, \ldots, u_N = \lambda_m(v)$$

in $\mathcal{U}_\infty(\widetilde{A})$ with $\|u_{j+1} - u_j\| < 2$ for each j. We can find k greater than both n and m and $v_0, v_1, \ldots, v_N$ in $\mathcal{U}_k(\widetilde{A})$ such that $\lambda_k(v_j) = u_j$. Now, $v_j \sim_h v_{j+1}$ in $\mathcal{U}_k(\widetilde{A})$ by Lemma 2.1.3 (iii), $v_0 = u \oplus 1_{k-n}$, and $v_1 = v \oplus 1_{k-m}$. This shows that $u \sim_1 v$. Conversely, if $u \sim_1 v$, then $u \oplus 1_{k-n} \sim_h v \oplus 1_{k-m}$ in $\mathcal{U}_k(\widetilde{A})$ for some k, and hence

$$\lambda_n(u) = \lambda_k(u \oplus 1_{k-n}) \sim_h \lambda_k(v \oplus 1_{k-m}) = \lambda_m(v)$$

in $\mathcal{U}_\infty(\tilde{A})$.

It follows that $K_1(A)$ is (isomorphic to) $\mathcal{U}_\infty(\tilde{A})/\sim_h$ with the new definition of $\mathcal{U}_\infty(\tilde{A})$. The inclusion $\mathcal{V}(A) \to \mathcal{U}_\infty(\tilde{A})$ is a π_0-equivalence, see Paragraph 11.2.1. This can be seen using the retract $\mathcal{U}_\infty(\tilde{A}) \to \mathcal{V}(A)$ given by $u \mapsto s(u)^*u$. Consequently,

$$\pi_0(\mathcal{V}(A)) = \mathcal{V}(A)/\sim_h \cong \mathcal{U}_\infty(\tilde{A})/\sim_h = K_1(A).$$

For $n \geqslant 1$ identify S^n with the quotient space $[0,1]^n/\partial[0,1]^n$, where the boundary $\partial[0,1]^n$ of $[0,1]^n$ is identified with a single point. Let $\Gamma_n(A)$ be the set of all continuous functions $f\colon [0,1]^n \to \mathcal{V}(A)$ with $f(t) = 1$ for all t in $\partial[0,1]^n$. Then $\Gamma_n(A)$ is a locally path-wise connected topological group, and $\pi_n(\mathcal{V}(A))$ is (isomorphic to) $\Gamma_n(A)/\sim_h$. Since

$$\Gamma_1(A) = \{u \in C([0,1], \mathcal{V}(A)) : u(0) = u(1) = 1\}$$
$$= \{u \in \mathcal{U}_\infty(\widetilde{SA}) : s(u) = 1\} = \mathcal{V}(SA),$$

we find that

$$\pi_1(\mathcal{V}(A)) = \Gamma_1(A)/\sim_h = \mathcal{V}(SA)/\sim_h \cong K_1(SA) = K_2(A).$$

Identify $\mathcal{V}(SA)$ with the set of all continuous functions $u\colon [0,1] \to \mathcal{U}_\infty(\tilde{A})$ satisfying $u(0) = u(1) = 1 = s(u(t))$ for each t in $[0,1]$. For each $n \geqslant 2$ identify $\Gamma_n(A)$ with $\Gamma_{n-1}(SA)$ as follows: given f in $\Gamma_n(A)$ and $(t_1, t_2, \ldots, t_{n-1})$ in $[0,1]^{n-1}$, define $\overline{f}(t_1, t_2, \ldots, t_{n-1})$ in $\mathcal{V}(SA)$ by

$$\overline{f}(t_1, t_2, \ldots, t_{n-1})(t) = f(t_1, t_2, \ldots, t_{n-1}, t).$$

Then $\overline{f}$ belongs to $\Gamma_{n-1}(SA)$, and the map $f \mapsto \overline{f}$ is an isomorphism of topological groups. It now follows by induction that

$$\pi_n(\mathcal{V}(A)) = \Gamma_n(A)/\sim_h \cong \Gamma_{n-1}(SA)/\sim_h \cong K_n(SA) = K_{n+1}(A)$$

for all $n \geqslant 2$.

Equation (11.10) is an immediate consequence of the Bott periodicity theorem (see Corollary 11.3.1). $\quad\square$

11.5* The holomorphic function calculus

The holomorphic function calculus is an indispensable tool when dealing with Banach algebras, and in particular for the study of K-theory for Banach algebras. Since we are concerned with C^*-algebras only, the continuous function calculus has been sufficient for our purposes, except at one point: the proof of Lemma 11.2.9, where we want to pass from generalized idempotents to idempotents. Here we are forced to deal with elements which are not necessarily normal, and the holomorphic function calculus is required to handle this situation.

Our first definition sets out some terminology, well known from classical complex analysis.

Definition 11.5.1. An *integration path* is a continuous and piece-wise continuously differentiable map $\gamma \colon [a, b] \to \mathbb{C}$, where $-\infty < a < b < \infty$. We set $\operatorname{Im}\gamma = \gamma([a, b])$. If $\gamma(a) = \gamma(b)$, then γ is *closed*.

A finite collection $\Gamma = \{\gamma_1, \ldots, \gamma_n\}$ of pairwise disjoint, closed integration paths is a *contour*. We set $\operatorname{Im}\Gamma = \operatorname{Im}\gamma_1 \cup \cdots \cup \operatorname{Im}\gamma_n$ and define

$$\int_\Gamma f(z)\,dz = \int_{\gamma_1} f(z)\,dz + \cdots + \int_{\gamma_n} f(z)\,dz$$

for every piece-wise continuous function f defined on $\operatorname{Im}\Gamma$. For every z in $\mathbb{C} \setminus \operatorname{Im}\Gamma$, the *index* of z with respect to Γ is the integer given by

$$\operatorname{Ind}_\Gamma(z) = \frac{1}{2\pi i} \int_\Gamma \frac{d\zeta}{\zeta - z}.$$

Suppose that $K \subset \Omega \subseteq \mathbb{C}$. The contour Γ *surrounds* K *in* Ω if $\operatorname{Im}\Gamma$ is contained in $\Omega \setminus K$ and if

$$\operatorname{Ind}_\Gamma(z) = \begin{cases} 1, & z \in K, \\ 0, & z \in \mathbb{C} \setminus \Omega. \end{cases}$$

It is an elementary topological fact that, for every compact subset K of an open set Ω in $\mathbb{C}$, there exists a contour $\Gamma = \{\gamma_1, \ldots, \gamma_n\}$ surrounding K in Ω. In this case we may even take each of the integration paths $\gamma_1, \ldots, \gamma_n$ to consist of finitely many oriented line segments, each parallel to either the real or the imaginary axis.

11.5.2 Definition of the holomorphic function calculus. Let a be an element in a unital C^*-algebra A, let Ω be an open neighborhood of sp(a), and take a contour Γ surrounding sp(a) in Ω. Write $\mathcal{H}(\Omega)$ for the set of holomorphic functions $h\colon \Omega \to \mathbb{C}$. Using the theory of Riemann integrals of piece-wise continuous Banach space valued functions (see [26, p. 203]), we may define

$$(11.11) \qquad h(a) = \frac{1}{2\pi i}\int_\Gamma h(z)(z1_A - a)^{-1}\, dz$$

for every function h in $\mathcal{H}(\Omega)$, and $h(a)$ is an element in A. It follows from Cauchy's theorem [38, Theorem 3.31] that this definition is independent of the choice of Γ. The map $h \mapsto h(a)$ is called the *holomorphic function calculus* for a. Its main properties are as follows.

Theorem 11.5.3. *Let a be an element in a unital C^*-algebra A, and let Ω be an open neighborhood of* sp(a).

(i) *The map $h \mapsto h(a)$, $\mathcal{H}(\Omega) \to A$, is a unital algebra homomorphism.*

(ii) *Suppose that h is given by a power series with radius of convergence greater than the spectral radius, $r(a)$, of a, say $h(z) = \sum_{j=0}^{\infty}\alpha_j z^j$. Then the definition (11.11) above of $h(a)$ agrees with the usual definition, i.e.,*

$$\frac{1}{2\pi i}\int_\Gamma h(z)(z1_A - a)^{-1}\, dz = \sum_{j=0}^{\infty}\alpha_j a^j.$$

(iii) *(The spectral mapping theorem) For every function h in $\mathcal{H}(\Omega)$, we have that* sp($h(a)$) $=$ $h($sp(a)$)$.

(iv) *Suppose that $\{h_n\}_{n=1}^{\infty}$ is a sequence of functions in $\mathcal{H}(\Omega)$ which converges uniformly on each compact subset of Ω to a function h. Then h belongs to $\mathcal{H}(\Omega)$, and $\{h_n(a)\}_{n=1}^{\infty}$ converges to $h(a)$ in norm.*

The proofs of the above results can be found in [26, §3.3]. For further results about the holomorphic function calculus and various extensions of it, we refer to [14, Chapter 2].

11.6 Exercises

Exercise 11.1. Let A be a C^*-algebra. By Bott periodicity (see Corollary 11.3.1) there exist group isomorphisms $\alpha_0 \colon K_0(A) \to K_0(S^2 A)$ and $\alpha_1 \colon K_1(A) \to K_1(S^2 A)$.

   (i) Express α_0 and α_1 in terms of the isomorphisms β_D and θ_E for suitable C^*-algebras D and E.

   (ii) Show that for each projection p in $\mathcal{P}(A)$ there is a projection q in $\mathcal{P}_2(\widetilde{S^2 A})$ such that $\alpha_0([p]_0) = [q]_0 - [s(q)]_0$. [Hint: Use (i).]

   (iii) Assume that A is unital. Show that for each unitary u in $\mathcal{U}(A)$ there is a unitary v in $\mathcal{U}_2(\widetilde{S^2 A})$ such that $\alpha_1([u]_1) = [v]_1$. [Hint: Use (i).]

Exercise 11.2. Show that the map $\dim \colon K_0(C(S^n)) \to \mathbb{Z}$ is an isomorphism for n odd, and that $\dim$ fails to be injective when n is even.

   Find a projection p in $\mathcal{P}_2(C(S^2))$ such that $K_0(C(S^2)) = \mathbb{Z}[1]_0 + \mathbb{Z}[p]_0$.

Exercise 11.3. Show that $\mathcal{U}(C(S^n))$ is connected for every natural number $n \geqslant 2$. [Hint: In the language of homotopy theory, this amounts to proving that any pair of continuous maps $S^n \to S^1$ are homotopic when $n \geqslant 2$, i.e., that $\pi_n(S^1) = 0$. Show first that $\pi_n(S^1) \cong \pi_n(\mathbb{R})$ (using that $\mathbb{R}$ is the covering space of S^1, that $\pi_1(S^n) = 0$, and hence that each continuous map $S^n \to S^1$ lifts to a continuous map $S^n \to \mathbb{R}$). Show next that $\pi_n(\mathbb{R}) = 0$ (using that $\mathbb{R}$ is contractible).]

   Show that the map

$$\Delta \colon K_1(C(S^n)) \to \mathcal{U}(C(S^n))/\mathcal{U}_0(C(S^n))$$

is an isomorphism if $n = 1$ or if n is even, and that Δ fails to be injective if n is odd and $n \geqslant 3$.

   Show that there exists a unitary u in $\mathcal{U}_2(C(S^3))$ such that $K_1(C(S^3)) = \mathbb{Z}[u]_1$ and $\Delta(u) = 1$. [Hint: Use Exercise 11.1.]

Exercise 11.4. Let e be an idempotent element in a unital C^*-algebra A. Show that

   (i) $\mathrm{sp}(e) \subseteq \{0, 1\}$,

   (ii) $\mathrm{sp}(e) = \{0\}$ if and only if $e = 0$,

(iii) $\mathrm{sp}(e) = \{1\}$ if and only if $e = 1_A$.

Exercise 11.5. Let A be a unital C^*-algebra containing an element with disconnected spectrum. Show that A contains a non-trivial projection, i.e., a projection different from 0 and 1_A. (See Exercise 1.5.)

Let A be a non-unital C^*-algebra containing an element with disconnected spectrum. Show that A contains a non-zero projection.

Exercise 11.6. Consider the nilpotent element

$$a = \begin{pmatrix} 0 & 1 \\ 0 & 0 \end{pmatrix} \in M_2(\mathbb{C}),$$

and observe that $\mathrm{sp}(a) = \{0\}$. Let Ω be an open neighborhood of 0 in $\mathbb{C}$, and let f be in $\mathcal{H}(\Omega)$. Show that

$$f(a) = \begin{pmatrix} f(0) & f'(0) \\ 0 & f(0) \end{pmatrix}.$$

[Hint: Use Theorem 11.5.3 (ii)].

Chapter 12

The Six-Term Exact Sequence

Combining Bott periodicity with the long exact sequence in K-theory we obtain the six-term exact sequence which will be the culmination of this introductory treatment of K-theory for C^*-algebras. With the six-term exact sequence at hand it is possible to calculate the K-theory of quite a few C^*-algebras as is illustrated in the examples and exercises of Chapters 11–13.

K-theory for C^*-algebras extends far beyond the treatment of this book, and there are theorems available that allow you to calculate the K-theory of crossed products (by $\mathbb{Z}$ and $\mathbb{R}$), tensor products (of "nice" C^*-algebras), (certain) group C^*-algebras, and (certain) reduced free product C^*-algebras, and the list does not stop here. These results are, unfortunately, beyond the scope of this book.

We already know five of the six maps going into the six-term exact sequence. The last map is the so-called exponential map, which is the Bott map composed with the index map δ_2.

12.1 The exponential map and the six-term exact sequence

12.1.1 The exponential map. For every short exact sequence of C^*-algebras

$$0 \longrightarrow I \xrightarrow{\varphi} A \xrightarrow{\psi} B \longrightarrow 0$$

define the *exponential map* $\delta_0 \colon K_0(B) \to K_1(I)$ to be the composition of the maps

$$K_0(B) \xrightarrow{\beta_B} K_2(B) \xrightarrow{\delta_2} K_1(I),$$

209

where δ_2 is the index map defined in Paragraph 10.2.3. In other words, if $\overline{\delta}_1$ is the index map associated with the suspended short exact sequence

$$0 \longrightarrow SI \xrightarrow{S\varphi} SA \xrightarrow{S\psi} SB \longrightarrow 0,$$

then

(12.1)
$$
\begin{array}{ccc}
K_0(B) & \xrightarrow{\delta_0} & K_1(I) \\
\downarrow{\scriptstyle \beta_B} & & \downarrow{\scriptstyle \theta_I} \\
K_1(SB) & \xrightarrow{\overline{\delta}_1} & K_0(SI)
\end{array}
$$

is a commutative diagram.

Theorem 12.1.2 (The six-term exact sequence). *For every short exact sequence of C^*-algebras*

(12.2)
$$0 \longrightarrow I \xrightarrow{\varphi} A \xrightarrow{\psi} B \longrightarrow 0,$$

the associated six-term sequence

(12.3)
$$
\begin{array}{ccccc}
K_0(I) & \xrightarrow{K_0(\varphi)} & K_0(A) & \xrightarrow{K_0(\psi)} & K_0(B) \\
\uparrow{\scriptstyle \delta_1} & & & & \downarrow{\scriptstyle \delta_0} \\
K_1(B) & \xleftarrow{K_1(\psi)} & K_1(A) & \xleftarrow{K_1(\varphi)} & K_1(I)
\end{array}
$$

is exact.

Proof. By Propositions 10.2.4 and 9.3.3 it remains to prove that (12.3) is exact at $K_0(B)$ and at $K_1(I)$. To see this consider the diagram

$$
\begin{array}{ccccccc}
K_2(A) & \xrightarrow{K_2(\psi)} & K_2(B) & \xrightarrow{\delta_2} & K_1(I) & \xrightarrow{K_1(\varphi)} & K_1(A) \\
\uparrow{\scriptstyle \beta_A}\cong & & \uparrow{\scriptstyle \beta_B}\cong & & \| & & \| \\
K_0(A) & \xrightarrow{K_0(\psi)} & K_0(B) & \xrightarrow{\delta_0} & K_1(I) & \xrightarrow{K_1(\varphi)} & K_1(A).
\end{array}
$$

The diagram is commutative: The left-hand square commutes by naturality of the Bott map, and the center square commutes by the definition of the exponential map. The top row is exact by Proposition 10.2.4, and this entails that the bottom row is exact. □

12.2 An explicit description of the exponential map

The exponential map is composed of two natural transformations and is therefore itself natural. This is the content of the following result.

Proposition 12.2.1. *The exponential map δ_0 is natural in the following sense. Given a commutative diagram*

(12.4)

$$
\begin{array}{ccccccccc}
0 & \longrightarrow & I & \overset{\varphi}{\longrightarrow} & A & \overset{\psi}{\longrightarrow} & B & \longrightarrow & 0 \\
& & \gamma \downarrow & & \alpha \downarrow & & \downarrow \beta & & \\
0 & \longrightarrow & I' & \underset{\varphi'}{\longrightarrow} & A' & \underset{\psi'}{\longrightarrow} & B' & \longrightarrow & 0
\end{array}
$$

with short exact rows of C^*-algebras, and where α, β, γ are $*$-homomorphisms, the diagram

(12.5)

$$
\begin{array}{ccc}
K_0(B) & \overset{\delta_0}{\longrightarrow} & K_1(I) \\
K_0(\beta) \downarrow & & \downarrow K_1(\gamma) \\
K_0(B') & \underset{\delta_0'}{\longrightarrow} & K_1(I')
\end{array}
$$

is commutative, where δ_0 and δ_0' are the associated exponential maps.

Proof. The diagram (12.5) can be decomposed into two commuting squares:

$$
\begin{array}{ccccc}
K_0(B) & \overset{\beta_B}{\longrightarrow} & K_2(B) & \overset{\delta_2}{\longrightarrow} & K_1(I) \\
K_0(\beta) \downarrow & & K_2(\beta) \downarrow & & \downarrow K_1(\gamma) \\
K_0(B') & \underset{\beta_{B'}}{\longrightarrow} & K_2(B') & \underset{\delta_2'}{\longrightarrow} & K_1(I'),
\end{array}
$$

see (11.1) and (10.5). □

The somewhat technical proposition below gives an explicit description of the exponential map, and it also justifies its name.

Proposition 12.2.2. *Let*

$$0 \longrightarrow I \overset{\varphi}{\longrightarrow} A \overset{\psi}{\longrightarrow} B \longrightarrow 0$$

be a short exact sequence of C^*-algebras, and let $\delta_0 \colon K_0(B) \to K_1(I)$ be its associated exponential map. Let g be an element in $K_0(B)$. Then $\delta_0(g)$ can be calculated as follows.

(i) *Let p be a projection in $\mathcal{P}_n(\widetilde{B})$ such that $g = [p]_0 - [s(p)]_0$, and let a be a self-adjoint element in $M_n(\widetilde{A})$ for which $\widetilde{\psi}(a) = p$. Then $\widetilde{\varphi}(u) = \exp(2\pi i a)$ for precisely one unitary u in $\mathcal{U}_n(\widetilde{I})$, and $\delta_0(g) = -[u]_1$.*

(ii) *Suppose that A is unital, in which case also B is unital and ψ is unit preserving. Let $\overline{\varphi} \colon \widetilde{I} \to A$ be given by $\overline{\varphi}(x + \alpha 1) = \varphi(x) + \alpha 1_A$. Suppose that $g = [p]_0$ for some projection p in $\mathcal{P}_n(B)$, and let a be a self-adjoint element in $M_n(A)$ for which $\psi(a) = p$. Then $\overline{\varphi}(u) = \exp(2\pi i a)$ for precisely one unitary element u in $\mathcal{U}_n(\widetilde{I})$, and $\delta_0(g) = -[u]_1$.*

By the standard picture for K_0 (Proposition 4.2.2) we can find a projection p as in (i) for every element g in $K_0(B)$, whereas only elements g in $K_0(B)^+$ can be represented by a projection p as in (ii). However, when B is unital, any element of $K_0(B)$ is the *difference* of two elements in $K_0(B)^+$. We can always find a self-adjoint element a that lifts p as in (i) and (ii), see Paragraph 2.2.10.

Proof of proposition.   (ii). We assume in the proof that B is non-zero, and hence that the map $\overline{\varphi} \colon M_n(\widetilde{I}) \to M_n(A)$ is injective for each natural number n. We leave it to the reader to consider the case where $B = 0$.

The image of $\overline{\varphi} \colon M_n(\widetilde{I}) \to M_n(A)$ consists of those elements x in $M_n(A)$ for which $\psi(x)$ belongs to $M_n(\mathbb{C}1_B)$. Because

$$\psi(\exp(2\pi i a)) = \exp(2\pi i \psi(a)) = \exp(2\pi i p) = 1,$$

there exists an element u in $M_n(\widetilde{I})$ such that $\overline{\varphi}(u) = \exp(2\pi i a)$, and since $\overline{\varphi}(u)$ is unitary, we conclude that u is unitary. By (12.1) we must show that

(12.6) $$(\overline{\delta}_1 \circ \beta_B)([p]_0) = \theta_I([u^*]_1),$$

where $\bar{\delta}_1 \colon K_1(SB) \to K_0(SI)$ is the index map associated with the short exact sequence

$$0 \longrightarrow SI \xrightarrow{\;S\varphi\;} SA \xrightarrow{\;S\psi\;} SB \longrightarrow 0.$$

We use here the picture $SB = C_0((0,1), B)$, in which $M_k(\widetilde{SB})$ is the set of all continuous functions $f \colon [0,1] \to M_k(B)$, where $f(0) = f(1)$ belongs to $M_k(\mathbb{C}1_B)$, and the projection loop f_p in $\mathcal{U}_n(\widetilde{SB})$ associated with the projection p is given by

$$f_p(t) = e^{2\pi it}p + (1-p) = e^{2\pi itp}, \quad t \in [0,1].$$

Let v in $\mathcal{U}_{2n}(\widetilde{SA})$ be such that $\widetilde{S\psi}(v) = \operatorname{diag}(f_p, f_p^*)$. Then $v \colon [0,1] \to \mathcal{U}_{2n}(A)$ is a continuous map with $v(0) = v(1)$ belonging to $M_{2n}(\mathbb{C}1_A)$, and

$$\psi(v(t)) = \begin{pmatrix} f_p(t) & 0 \\ 0 & f_p(t)^* \end{pmatrix}, \quad t \in [0,1].$$

As $f_p(0) = f_p(1) = 1$ we have $v(0) = v(1) = 1$.

With a being the self-adjoint lift in $M_n(A)$ of p, put $z(t) = \exp(2\pi ita)$ for t in $[0,1]$. Then $z(t)$ belongs to $\mathcal{U}_n(A)$, the map $t \mapsto z(t)$ is continuous, and $\psi(z(t)) = f_p(t)$. Hence

$$\psi\!\left(v(t) \begin{pmatrix} z(t)^* & 0 \\ 0 & z(t) \end{pmatrix}\right) = 1_{2n}, \qquad s\!\left(v(t) \begin{pmatrix} z(t)^* & 0 \\ 0 & z(t) \end{pmatrix}\right) = 1_{2n}.$$

It follows that we can find $w(t)$ in $\mathcal{U}_{2n}(\widetilde{I})$ with

$$\overline{\varphi}(w(t)) = v(t) \begin{pmatrix} z(t)^* & 0 \\ 0 & z(t) \end{pmatrix}, \qquad s(w(t)) = 1_{2n}.$$

Now, $t \mapsto w(t)$ is continuous because $\overline{\varphi}$ is isometric, $w(0) = 1$, and

$$\overline{\varphi}(w(1)) = \begin{pmatrix} z(1)^* & 0 \\ 0 & z(1) \end{pmatrix} = \overline{\varphi} \begin{pmatrix} u^* & 0 \\ 0 & u \end{pmatrix},$$

which shows that $w(1) = \operatorname{diag}(u^*, u)$. Theorem 10.1.3 yields

$$(12.7) \qquad \theta_I([u^*]_1) = \left[w \begin{pmatrix} 1_n & 0 \\ 0 & 0 \end{pmatrix} w^* \right]_0 - \left[\begin{pmatrix} 1_n & 0 \\ 0 & 0 \end{pmatrix} \right]_0.$$

We also have

$$\overline{\varphi}(w(t) \begin{pmatrix} 1_n & 0 \\ 0 & 0 \end{pmatrix} w(t)^*) = v(t) \begin{pmatrix} 1_n & 0 \\ 0 & 0 \end{pmatrix} v(t)^*,$$

which implies that

(12.8) $$\widetilde{S\varphi}(w \begin{pmatrix} 1_n & 0 \\ 0 & 0 \end{pmatrix} w^*) = v \begin{pmatrix} 1_n & 0 \\ 0 & 0 \end{pmatrix} v^*.$$

The unitary v was chosen such that $\widetilde{S\psi}(v) = \mathrm{diag}(f_p, f_p^*)$, and so we get from equations (12.8) and (12.7) and the definition of the index map (Definition 9.1.3) that $\overline{\delta}_1([f_p]_1) = \theta_I([u^*]_1)$. This, combined with the identity $\beta_B([p]_0) = [f_p]_1$, proves (12.6).

(i). Consider the diagram

$$
\begin{array}{ccccccccc}
0 & \longrightarrow & I & \xrightarrow{\varphi} & A & \xrightarrow{\psi} & B & \longrightarrow & 0 \\
 & & \Vert & & \downarrow{\scriptstyle \iota_A} & & \downarrow{\scriptstyle \iota_B} & & \\
0 & \longrightarrow & I & \xrightarrow{\varphi'} & \widetilde{A} & \xrightarrow{\psi'} & \widetilde{B} & \longrightarrow & 0,
\end{array}
$$

where $\varphi' = \iota_A \circ \varphi$ and $\psi' = \widetilde{\psi}$. Let δ_0' be the exponential map associated with the short exact sequence in its lower row. By naturality of the exponential map (Proposition 12.2.1),

$$\delta_0([p]_0 - [s(p)]_0) = (\delta_0' \circ K_0(\iota_B))([p]_0 - [s(p)]_0) = \delta_0'([p]_0 - [s(p)]_0)$$
$$= \delta_0'([p]_0) - \delta_0'([s(p)]_0).$$

The maps $\widetilde{\varphi} \colon \widetilde{I} \to \widetilde{A}$ and $\overline{\varphi'} \colon \widetilde{I} \to \widetilde{A}$ coincide. It follows from (ii) that there is a unitary u in $\mathcal{U}_n(\widetilde{I})$ such that $\widetilde{\varphi}(u) = \overline{\varphi'}(u) = \exp(2\pi i a)$ and that $\delta_0'([p]_0) = -[u]_1$. Since $s(p) = \psi'(s(p))$, with $s(p)$ viewed as a scalar projection belonging to $M_n(\widetilde{B})$ as well as to $M_n(\widetilde{A})$, it follows that $[s(p)]_0$ belongs to the image of $K_0(\psi')$ and hence to the kernel of δ_0'. This completes the proof. $\qquad\square$

12.3 Exercises

Exercise 12.1. Show that K_1 is split exact (Proposition 8.2.5) using Theorem 12.1.2 and half exactness of K_1 (Proposition 8.2.4).

Exercise 12.2. Let n be a natural number. Consider the equivalence relation $\sim_n$ on the closed unit disc $\mathbb{D}$ given by $z_1 \sim_n z_2$ when $|z_1| = |z_2| = 1$ and $z_1^n = z_2^n$, or when $z_1 = z_2$. Let X_n be the quotient space $\mathbb{D}/\sim_n$. The space X_n is called a *Moore space*.

(i) Put $\omega = \exp(2\pi i/n)$. Show that

$$C(X_n) = \{f \in C(\mathbb{D}) : f(\omega z) = f(z), \ z \in \mathbb{T}\}.$$

(ii) Let v be the unitary element in $C(\mathbb{T})$ given by $v(z) = z$. Recall from Example 11.3.4 that $[v]_1$ is a generator for the cyclic group $K_1(C(\mathbb{T}))$. Consider the *-homomorphism $\rho_n \colon C(\mathbb{T}) \to C(\mathbb{T})$ given by $(\rho_n(f))(z) = f(z^n)$. Find a *-homomorphism π making the following diagram commutative with exact rows:

$$
\begin{array}{ccccccccc}
0 & \longrightarrow & C_0(\mathbb{R}^2) & \overset{\varphi}{\longrightarrow} & C(X_n) & \overset{\pi}{\longrightarrow} & C(\mathbb{T}) & \longrightarrow & 0 \\
 & & \| & & \iota\uparrow & & \downarrow{\rho_n} & & \\
0 & \longrightarrow & C_0(\mathbb{R}^2) & \underset{\varphi}{\longrightarrow} & C(\mathbb{D}) & \underset{\psi}{\longrightarrow} & C(\mathbb{T}) & \longrightarrow & 0.
\end{array}
$$

The bottom row is the short exact sequence from Exercise 2.12 and ι is the inclusion mapping.

(iii) Show that the index map $\delta_1 \colon K_1(C(\mathbb{T})) \to K_0(C_0(\mathbb{R}^2))$ associated with the short exact sequence in (ii) maps a generator of the group $K_1(C(\mathbb{T}))$ to n times a generator of the group $K_0(C_0(\mathbb{R}^2))$.

(iv) Show that $K_0(C(X_n)) \cong \mathbb{Z} \oplus (\mathbb{Z}/n\mathbb{Z})$, and that $K_1(C(X_n)) = 0$.

Exercise 12.3. For each natural number n let Z_n be the n-clover obtained by forming the disjoint union of n circles, choosing one point in each circle, and then identifying these n points, so that they become one common point. In particular, $Z_1 = \mathbb{T}$ and Z_2 is the "figure eight".
    Calculate $K_0(C(Z_n))$ and $K_1(C(Z_n))$ for each n by constructing a short

exact sequence

$$0 \longrightarrow C_0(U) \longrightarrow C(Z_n) \longrightarrow C(Y) \longrightarrow 0$$

for suitable spaces U and Y.

Exercise 12.4. Describe the six-term exact sequence (six groups and six group homomorphisms) associated with each of the following short exact sequences

(i) The exact sequence from Exercise 1.2:

$$0 \longrightarrow C_0((0,1)) \stackrel{\iota}{\longrightarrow} C([0,1]) \stackrel{\psi}{\longrightarrow} \mathbb{C} \oplus \mathbb{C} \longrightarrow 0,$$

where $\psi(f) = (f(0), f(1))$.

(ii) The exact sequence arising from adjoining a unit:

$$0 \longrightarrow A \stackrel{\iota}{\longrightarrow} \tilde{A} \stackrel{\pi}{\longrightarrow} \mathbb{C} \longrightarrow 0.$$

(iii) The exact sequence from Exercises 9.5 and 2.12:

$$0 \longrightarrow C_0(\mathbb{R}^2) \stackrel{\varphi}{\longrightarrow} C(\mathbb{D}) \stackrel{\psi}{\longrightarrow} C(\mathbb{T}) \longrightarrow 0,$$

where ψ is the restriction mapping, and where $\mathbb{R}^2$ is identified with $\mathbb{D}\backslash\mathbb{T}$.

(iv) The exact sequence from Example 9.4.3 and Formula (9.10):

$$0 \longrightarrow \mathcal{K} \stackrel{\iota}{\longrightarrow} B(H) \stackrel{\pi}{\longrightarrow} \mathcal{Q}(H) \longrightarrow 0,$$

where H is an infinite dimensional separable Hilbert space.

(v) The "suspension-cone extension" from Example 4.1.5:

$$0 \longrightarrow SA \stackrel{\iota}{\longrightarrow} CA \stackrel{\psi}{\longrightarrow} A \longrightarrow 0,$$

where $\psi(f) = f(1)$.

(vi) The Toeplitz algebra extension from Example 9.4.4:

$$0 \longrightarrow \mathcal{K} \stackrel{\iota}{\longrightarrow} \mathcal{T} \stackrel{\psi}{\longrightarrow} C(\mathbb{T}) \longrightarrow 0.$$

(vii) The reduced Toeplitz algebra extension from Example 9.4.4:

$$0 \longrightarrow \mathcal{K} \xrightarrow{\iota} \mathcal{T}_0 \xrightarrow{\psi} C_0(\mathbb{T} \setminus \{1\}) \longrightarrow 0.$$

Exercise 12.5. Let A and B be C^*-algebras, and let $\varphi \colon A \to B$ be a $*$-homomorphism. Let E_φ be the sub-C^*-algebra of $C([0,1], B) \oplus A$ consisting of those pairs (f, a) such that $f(0) = 0$ and $f(1) = \varphi(a)$. Let $\pi \colon E_\varphi \to A$ be given by $\pi(f, a) = a$.

Show that

$$0 \longrightarrow SB \xrightarrow{\iota} E_\varphi \xrightarrow{\pi} A \longrightarrow 0$$

is a short exact sequence of C^*-algebras. Let δ_0 and δ_1 be the exponential map and the index map associated with this sequence. Show that $\delta_0 = -\beta_B \circ K_0(\varphi)$ and that $\delta_1 = \theta_B \circ K_1(\varphi)$. Use this to show that the sequence

$$
\begin{array}{ccccc}
K_1(B) & \xrightarrow{\gamma_1} & K_0(E_\varphi) & \xrightarrow{K_0(\pi)} & K_0(A) \\
{\scriptstyle K_1(\varphi)} \big\uparrow & & & & \big\downarrow {\scriptstyle K_0(\varphi)} \\
K_1(A) & \xleftarrow[K_1(\pi)]{} & K_1(E_\varphi) & \xleftarrow[\gamma_0]{} & K_0(B)
\end{array}
$$

is exact, where $\gamma_0 = \delta_0 \circ \beta_B$ and $\gamma_1 = \delta_1 \circ \theta_B$.

Chapter 13

Inductive Limits of Dimension Drop Algebras

The purpose of this last chapter is twofold. First, the techniques developed in the previous chapters are used to calculate the K-theory of a natural and, with our techniques, manageable class of C^*-algebras. Secondly, we apply these results to show that for every pair of countable Abelian groups G_0 and G_1 there is a separable C^*-algebra A such that $K_0(A) \cong G_0$ and $K_1(A) \cong G_1$.

13.1   Dimension drop algebras

For each integer $n \geqslant 2$, the *dimension drop algebra* D_n is defined to be the C^*-algebra consisting of all functions f in $C([0, 1], M_n(\mathbb{C}))$ such that $f(0) = 0$ and $f(1)$ belongs to $\mathbb{C}1_n$. Since

$$SM_n(\mathbb{C}) = \{f \in C([0, 1], M_n(\mathbb{C})) : f(0) = f(1) = 0\} \subseteq D_n,$$

we obtain a short exact sequence

$$0 \longrightarrow SM_n(\mathbb{C}) \overset{\iota}{\longrightarrow} D_n \overset{\pi}{\longrightarrow} \mathbb{C} \longrightarrow 0$$

where $\pi(f) = f(1)$ (identifying $\mathbb{C}1_n$ with $\mathbb{C}$). This sequence gives rise to the six-term exact sequence

$$
\begin{array}{ccc}
K_0(SM_n(\mathbb{C})) \longrightarrow K_0(D_n) \longrightarrow K_0(\mathbb{C}) \\
\delta_1 \Big\uparrow \qquad\qquad\qquad\qquad\qquad \Big\downarrow \delta_0 \\
K_1(\mathbb{C}) \longleftarrow K_1(D_n) \longleftarrow K_1(SM_n(\mathbb{C})).
\end{array}
$$

Since $K_0(SM_n(\mathbb{C})) \cong K_1(M_n(\mathbb{C})) = 0$ and $K_1(\mathbb{C}) = 0$, the six-term exact sequence reduces to the exact sequence

$$0 \longrightarrow K_0(D_n) \xrightarrow{K_0(\pi)} K_0(\mathbb{C}) \xrightarrow{\delta_0} K_1(SM_n(\mathbb{C})) \xrightarrow{K_1(\iota)} K_1(D_n) \longrightarrow 0$$
$$\cong \downarrow \qquad\qquad\qquad\qquad \downarrow \cong$$
$$\mathbb{Z} \xrightarrow{\qquad n \qquad} \mathbb{Z}.$$

As indicated in the diagram, $K_0(\mathbb{C})$ and $K_1(SM_n(\mathbb{C}))$ are isomorphic to $\mathbb{Z}$. We proceed to show that the homomorphism $\mathbb{Z} \to \mathbb{Z}$ induced by δ_0 is multiplication by n. For this purpose identify $\widetilde{D}_n$ with the set of functions f in $C([0,1], M_n(\mathbb{C}))$ such that $f(0)$ and $f(1)$ belong to $\mathbb{C}1_n$, and identify $(SM_n(\mathbb{C}))^\sim$ with the set of functions f in $C([0,1], M_n(\mathbb{C}))$ where $f(0) = f(1)$ belongs to $\mathbb{C}1_n$. Let p be a fixed one-dimensional projection in $M_n(\mathbb{C})$. Then $[p]_0$ is a generator of the group $K_0(M_n(\mathbb{C}))$, see Example 3.3.2. Let u_n, v_n in $\mathcal{U}((SM_n(\mathbb{C}))^\sim) \subseteq \mathcal{U}(\widetilde{D}_n)$ be given by

(13.1)
$$u_n(t) = e^{2\pi i t}1_n, \qquad\qquad\qquad t \in [0, 2\pi],$$
$$v_n(t) = \exp(2\pi i t p) = e^{2\pi i t}p + (1_n - p), \qquad t \in [0, 2\pi].$$

By the definition of the Bott isomorphism $\beta\colon K_0(M_n(\mathbb{C})) \to K_1(SM_n(\mathbb{C}))$ from Chapter 11, $\beta([p]_0) = [v_n]_1$ and $\beta([1_n]_0) = [u_n]_1$. We conclude that $n[v_n]_1 = [u_n]_1$ in $K_1(SM_n(\mathbb{C}))$ and in $K_1(D_n)$, and that $[v_n]_1$ is a generator of the group $K_1(SM_n(\mathbb{C}))$ and of the group $K_1(D_n)$.

Let h in D_n be the self-adjoint element given by $h(t) = t \cdot 1_n$ for t in $[0,1]$. Then $\pi(h) = 1_\mathbb{C}$ and $u_n = \exp(2\pi i h)$, which by Proposition 12.2.2 (ii) entails that $\delta_0([1_\mathbb{C}]_0) = -[u_n]_1 = -n[v_n]_1$. This proves that δ_0 is injective and is multiplication by n (when $K_0(\mathbb{C})$ and $K_1(SM_n(\mathbb{C}))$ are identified with $\mathbb{Z}$). Hence $K_0(D_n) = 0$ and the image of δ_0 is generated by $n[v_n]_1$. Consequently, $[v_n]_1$ has order n in $K_1(D_n)$, and $K_1(D_n)$ is isomorphic to $\mathbb{Z}/n\mathbb{Z}$.

13.1.1 The building blocks. Let $\mathcal{B}_0$ be the class consisting of the C^*-algebras $C_0(\mathbb{R})$, $C_0(\mathbb{R}^2)$, D_n, and SD_n for each $n \geqslant 2$. The K-theory for these C^*-algebras is given by

	K_0	K_1
D_n	0	$\mathbb{Z}/n\mathbb{Z}$
SD_n	$\mathbb{Z}/n\mathbb{Z}$	0
$C_0(\mathbb{R})$	0	$\mathbb{Z}$
$C_0(\mathbb{R}^2)$	$\mathbb{Z}$	0

Notice that $C_0(\mathbb{R})$ is isomorphic to $C_0((0,1)) = S\mathbb{C}$, and that $C_0(\mathbb{R}^2)$ is isomorphic to $SC_0((0,1))$. Define v_1 in $\mathcal{U}(C_0((0,1))^\sim)$ by

$$(13.2) \qquad v_1(t) = e^{2\pi i t}, \quad t \epsilon [0,1].$$

Then $[v_1]_1 = \beta_{\mathbb{C}}([1_{\mathbb{C}}]_0)$, and therefore the group $K_1(C_0((0,1)))$ is generated by $[v_1]_1$.

Let $\mathcal{B}$ be the class of all C^*-algebras of the form

$$M_{k_1}(B_1) \oplus M_{k_2}(B_2) \oplus \cdots \oplus M_{k_r}(B_r),$$

where $r, k_1, \ldots, k_r$ are integers, and where $B_1, \ldots, B_r$ belong to $\mathcal{B}_0$.

Recall the following standard fact from group theory.

Proposition 13.1.2. *Every finitely generated Abelian group is isomorphic to a group of the form*

$$\mathbb{Z}^k \oplus (\mathbb{Z}/n_1\mathbb{Z}) \oplus \cdots \oplus (\mathbb{Z}/n_r\mathbb{Z}),$$

for some non-negative integers k, r and some natural numbers $n_1, \ldots, n_r$.

From Paragraph 13.1.1, Proposition 13.1.2, Proposition 4.3.4 and Proposition 8.2.6 we obtain the following result.

Corollary 13.1.3. *For each pair G_0 and G_1 of finitely generated Abelian groups there is a C^*-algebra A in $\mathcal{B}$ satisfying $K_0(A) \cong G_0$ and $K_1(A) \cong G_1$.*

13.2 Realizing countable Abelian groups as K-groups

We show that an arbitrary pair of countable Abelian groups can be realized as the K-groups of a C^*-algebra which is the inductive limit of C^*-algebras from the class $\mathcal{B}$. The main step in the proof is to "lift" a pair of group homomorphisms $\alpha_0 \colon K_0(A) \to K_0(B)$ and $\alpha_1 \colon K_1(A) \to K_1(B)$, where A, B belong to $\mathcal{B}$, to a *-homomorphism from A to (a matrix algebra over) B.

For any C^*-algebra A and any positive integer k, let $\mu_k, \lambda_k \colon A \to M_k(A)$ be given by

$$(13.3) \qquad \mu_k(a) = \begin{pmatrix} a & 0 & \cdots & 0 \\ 0 & a & \cdots & 0 \\ \vdots & \vdots & \ddots & \vdots \\ 0 & 0 & \cdots & a \end{pmatrix}, \quad \lambda_k(a) = \begin{pmatrix} a & 0 & \cdots & 0 \\ 0 & 0 & \cdots & 0 \\ \vdots & \vdots & \ddots & \vdots \\ 0 & 0 & \cdots & 0 \end{pmatrix}, \quad a \in A.$$

Recall from Propositions 4.3.8 and 8.2.8 that $K_0(\lambda_k)$ and $K_1(\lambda_k)$ are isomorphisms, and notice that

$$(13.4) \qquad K_0(\mu_k) = k \cdot K_0(\lambda_k), \qquad K_1(\mu_k) = k \cdot K_1(\lambda_k).$$

Lemma 13.2.1. *Let A, B be C^*-algebras in $\mathcal{B}_0$, and let $\alpha_0 \colon K_0(A) \to K_0(B)$ and $\alpha_1 \colon K_1(A) \to K_1(B)$ be group homomorphisms. Then there exist a positive integer k and a *-homomorphism $\varphi \colon A \to M_k(B)$ making the diagram*

$$\begin{array}{ccc} K_i(A) & \xrightarrow{\quad K_i(\varphi) \quad} & K_i(M_k(B)) \\ & {\alpha_i} \searrow \qquad \nearrow {K_i(\lambda_k)} & \\ & K_i(B) & \end{array}$$

commutative for $i = 0, 1$.

Proof. Inspecting the possible K-groups of A and B one finds that either $\alpha_0 = 0$ or $\alpha_1 = 0$. If both $\alpha_0 = 0$ and $\alpha_1 = 0$, then let φ be the zero map. Assume that $\alpha_1 \neq 0$. Then $\alpha_0 = 0$ and there are the following three possibilities:

  (i) $A = B = C_0((0,1))$,

  (ii) $A = C_0((0,1))$ and $B = D_n$, where $n \geqslant 2$,

  (iii) $A = D_{n_1}$ and $B = D_{n_2}$, where $n_1, n_2 \geqslant 2$.

We examine these three possibilities one by one. Recall that $K_1(C_0((0,1)))$ is generated by $[v_1]_1$, where v_1 is defined in (13.2), and that $K_1(D_n)$ is generated by $[v_n]_1$, where v_n is defined in (13.1).

(i). There is a non-zero integer m such that $\alpha_1([v_1]_1) = m[v_1]_1$. Consider first the case where m is positive. Put $k = m$, and set $\varphi = \mu_k \colon C_0((0,1)) \to M_k(C_0((0,1)))$. Then

$$K_1(\varphi)\,([v_1]_1) = k \cdot K_1(\lambda_k)\,([v_1]_1) = (K_1(\lambda_k) \circ \alpha_1)\,([v_1]_1),$$

which shows that $K_1(\varphi) = K_1(\lambda_k) \circ \alpha_1$.

Assume next that m is negative. Let τ be the automorphism on $C_0((0,1))$ given by $(\tau(f))(t) = f(1-t)$ for t in $[0,1]$. Then $\tau(v_1) = v_1^*$, which implies that $K_1(\tau)([v_1]_1) = -[v_1]_1$. Put $k = -m$, and put $\varphi = \mu_k \circ \tau \colon C_0((0,1)) \to M_k(C_0((0,1)))$. Then

$$K_1(\varphi)([v_1]_1) = -k \cdot K_1(\lambda_k)([v_1]_1) = (K_1(\lambda_k) \circ \alpha_1)([v_1]_1),$$

and so $K_1(\varphi) = K_1(\lambda_k) \circ \alpha_1$.

(ii). There is k in $\{1, 2, \ldots, n-1\}$ such that $\alpha_1([v_1]_1) = k[v_n]_1$. Let φ be the composition of the *-homomorphisms

$$C_0((0,1)) \xrightarrow{\lambda_n} M_n(C_0((0,1))) \xrightarrow{\mu_k} M_k(M_n(C_0((0,1)))) \xhookrightarrow{\iota} M_k(D_n).$$

Then $\widetilde{\varphi}(v_1) = (\widetilde{\mu}_k \circ \widetilde{\lambda}_n)(v_1)$ is homotopic to $\widetilde{\mu}_k(v_n)$ in $\mathcal{U}_k(\widetilde{D}_n)$. By (13.4), $K_1(\varphi)([v_1]_1) = k \cdot K_1(\lambda_k)([v_n]_1)$, and this shows that $K_1(\varphi) = K_1(\lambda_k) \circ \alpha_1$.

(iii). Let m in $\{1, 2, \ldots, n_2 - 1\}$ be such that $\alpha_1([v_{n_1}]_1) = m[v_{n_2}]_1$. As $n_1[v_{n_1}]_1 = 0$, we can find a positive integer k such that $n_1 m = n_2 k$. Set $n = n_1 m = n_2 k$. The image of the *-homomorphism $\mu_m \colon D_{n_1} \to M_m(D_{n_1})$ is contained in D_n. There is a natural inclusion $\iota \colon D_n \to M_k(D_{n_2})$. Let φ be the composition of the *-homomorphisms

$$D_{n_1} \xrightarrow{\mu_m} D_n \xhookrightarrow{\iota} M_k(D_{n_2}).$$

By (13.4), $K_1(\varphi)([v_{n_1}]_1) = m \cdot K_1(\lambda_k)([v_{n_2}]_1)$, and so $K_1(\varphi) = K_1(\lambda_k) \circ \alpha_1$.

Assume now that $\alpha_0 \neq 0$ and $\alpha_1 = 0$. Then $A = SA'$ and $B = SB'$, where A' and B' are equal to $C_0((0,1))$ or to D_n. Identifying the C^*-algebras $SM_k(B')$ and $M_k(B)$, we can use the first part of the proof to find a *-homomorphism $\varphi \colon A' \to M_k(B')$ such that $S\varphi \colon A \to M_k(B)$ satisfies $K_0(S\varphi) = K_0(\lambda_k) \circ \alpha_0$. $\qquad\square$

Lemma 13.2.2. *Let C^*-algebras A in $\mathcal{B}$ and B in $\mathcal{B}_0$ be given, and let $\alpha_0\colon K_0(A) \to K_0(B)$ and $\alpha_1\colon K_1(A) \to K_1(B)$ be group homomorphisms. Then there exist a positive integer k and a $*$-homomorphism $\varphi\colon A \to M_k(B)$ such that the diagram*

$$
\begin{array}{ccc}
K_i(A) & \xrightarrow{\ \ K_i(\varphi)\ \ } & K_i(M_k(B)) \\
& \searrow{\scriptstyle \alpha_i} \quad \nearrow{\scriptstyle K_i(\lambda_k)} & \\
& K_i(B) &
\end{array}
$$

commutes for $i = 0, 1$.

Proof. By assumption,

$$A = M_{l_1}(A_1) \oplus M_{l_2}(A_2) \oplus \cdots \oplus M_{l_r}(A_r)$$

for some positive integers $r, l_1, \ldots, l_r$ and some C^*-algebras $A_1, \ldots, A_r$ in $\mathcal{B}_0$. For each C^*-algebra D and for each positive integer n, let $\lambda_{n,D} = \lambda_n$ be as defined in (13.3). Let $\iota_j\colon M_{l_j}(A_j) \to A$ and $\pi_j\colon A \to M_{l_j}(A_j)$ be the canonical $*$-homomorphisms associated with the direct sum. Use Lemma 13.2.1 to find positive integers $k_1, \ldots, k_r$ and $*$-homomorphisms $\varphi_j\colon A_j \to M_{k_j}(B)$ such that

$$K_i(\varphi_j) = K_i(\lambda_{k_j,B}) \circ \alpha_i \circ K_i(\iota_j \circ \lambda_{l_j,A_j})\colon K_i(A_j) \to K_i(M_{k_j}(B))$$

for $i = 0, 1$ and for $j = 1, \ldots, r$. Let $\widehat{\varphi}_j\colon M_{l_j}(A_j) \to M_{k_j l_j}(B)$ be the $*$-homomorphism induced by φ_j as in (1.5).

Set $k = \sum_{j=1}^r k_j l_j$, and let $\eta_j\colon M_{k_j l_j}(B) \to M_k(B)$ be mutually orthogonal $*$-homomorphisms (obtained by dividing $M_k(B)$ into block matrices of appropriate sizes) such that $K_i(\eta_j \circ \lambda_{k_j l_j,B}) = K_i(\lambda_{k,B})$. Put

$$\varphi = \sum_{j=1}^r \eta_j \circ \widehat{\varphi}_j \circ \pi_j\colon A \to M_k(B).$$

By the definitions of $\widehat{\varphi}_j$ and φ_j we have

$$
\begin{aligned}
K_i(\widehat{\varphi}_j \circ \lambda_{l_j,A_j}) &= K_i(\lambda_{l_j,M_{k_j}(B)} \circ \varphi_j) \\
&= K_i(\lambda_{l_j,M_{k_j}(B)} \circ \lambda_{k_j,B}) \circ \alpha_i \circ K_i(\iota_j \circ \lambda_{l_j,A_j}) \\
&= K_i(\lambda_{k_j l_j,B}) \circ \alpha_i \circ K_i(\iota_j \circ \lambda_{l_j,A_j}),
\end{aligned}
$$

and therefore $K_i(\widehat{\varphi}) = K_i(\lambda_{k_j l_j, B}) \circ \alpha_i \circ K_i(\iota_j)$ because $K_i(\lambda_{l_j, A_j})$ is an isomorphism.

By Lemma 3.2.7, $\sum_{j=1}^{r} K_0(\iota_j \circ \pi_j) = \mathrm{id}_{K_0(A)}$. A similar argument shows that $\sum_{j=1}^{r} K_1(\iota_j \circ \pi_j) = \mathrm{id}_{K_1(A)}$. Hence

$$K_i(\varphi) = \sum_{j=1}^{r} K_i(\eta_j) \circ K_i(\lambda_{k_j l_j, B}) \circ \alpha_i \circ K_i(\iota_j) \circ K_i(\pi_j)$$

$$= \sum_{j=1}^{r} K_i(\lambda_{k, B}) \circ \alpha_i \circ K_i(\iota_j \circ \pi_j) = K_i(\lambda_{k, B}) \circ \alpha_i.$$

$\square$

Proposition 13.2.3. *Let A be a C^*-algebra in $\mathcal{B}$, let G_0 and G_1 be finitely generated Abelian groups, and let $\alpha_0 \colon K_0(A) \to G_0$ and $\alpha_1 \colon K_1(A) \to G_1$ be group homomorphisms. Then there are a C^*-algebra B in $\mathcal{B}$, a *-homomorphism $\varphi \colon A \to B$, and group isomorphisms $\beta_0 \colon K_0(B) \to G_0$, $\beta_1 \colon K_1(B) \to G_1$, making the diagram*

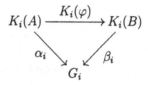

commutative for $i = 0, 1$.

Proof. By Paragraph 13.1.1 and Proposition 13.1.2 there are C^*-algebras $B_1, \ldots, B_r$ in $\mathcal{B}_0$ such that

$$G_i \cong K_i(B_1) \oplus K_i(B_2) \oplus \cdots \oplus K_i(B_r), \quad i = 0, 1.$$

Let $p_j^{(i)} \colon G_i \to K_i(B_j)$ and $q_j^{(i)} \colon K_i(B_j) \to G_i$ be the canonical group homomorphisms associated with the direct sum, so that

$$\mathrm{id}_{G_i} = \sum_{j=1}^{r} q_j^{(i)} \circ p_j^{(i)}, \quad \mathrm{id}_{K_i(B_j)} = p_j^{(i)} \circ q_j^{(i)}.$$

By Lemma 13.2.2 there are positive integers $k_1, \ldots, k_r$ and *-homomorphisms

$\varphi_j \colon A \to M_{k_j}(B_j)$ making the diagrams

$$
\begin{array}{ccc}
K_i(A) & \xrightarrow{\;\;K_i(\varphi_j)\;\;} & K_i(M_{k_j}(B_j)) \\
& & \\
{}_{p_j^{(i)} \circ \alpha_i} \searrow & & \nearrow {}_{K_i(\lambda_{k_j})} \\
& K_i(B_j) &
\end{array}
$$

commutative. Put

$$
B = M_{k_1}(B_1) \oplus M_{k_2}(B_2) \oplus \cdots \oplus M_{k_r}(B_r),
$$

and let $\pi_j \colon B \to M_{k_j}(B_j)$ and $\iota_j \colon M_{k_j}(B_j) \to B$ be the canonical *-homomor-phisms associated with the direct sum. Define the *-homomorphism $\varphi \colon A \to B$ by

$$
\varphi(a) = (\varphi_1(a), \varphi_2(a), \ldots, \varphi_r(a)), \quad a \in A.
$$

For $j = 1, \ldots, r$, let $\lambda_{k_j} \colon B_j \to M_{k_j}(B_j)$ be as defined in (13.3), and recall that $K_i(\lambda_{k_j})$ is an isomorphism. For $i = 0, 1$, define $\beta_i \colon K_i(B) \to G_i$ and $\gamma_i \colon G_i \to K_i(B)$ by

$$
\beta_i = \sum_{j=1}^{r} q_j^{(i)} \circ K_i(\lambda_{k_j})^{-1} \circ K_i(\pi_j), \qquad \gamma_i = \sum_{j=1}^{r} K_i(\iota_j) \circ K_i(\lambda_{k_j}) \circ p_j^{(i)}.
$$

Then $\beta_i \circ \gamma_i = \mathrm{id}_{G_i}$ and $\gamma_i \circ \beta_i = \mathrm{id}_{K_i(B)}$, and so β_i is an isomorphism. As $\pi_j \circ \varphi = \varphi_j$, we get

$$
\beta_i \circ K_i(\varphi) = \sum_{j=1}^{r} q_j^{(i)} \circ K_i(\lambda_{k_j})^{-1} \circ K_i(\varphi_j) = \sum_{j=1}^{r} q_j^{(i)} \circ p_j^{(i)} \circ \alpha_i = \alpha_i,
$$

and this completes the proof. $\square$

Theorem 13.2.4. *Let G_0 and G_1 be countable Abelian groups. Then there exists a C^*-algebra A such that $K_0(A) \cong G_0$ and $K_1(A) \cong G_1$. The C^*-algebra A can be chosen to be the inductive limit of a sequence*

$$
A_1 \xrightarrow{\;\varphi_1\;} A_2 \xrightarrow{\;\varphi_2\;} A_3 \xrightarrow{\;\varphi_3\;} \cdots,
$$

where $A_1, A_2, A_3, \ldots$ belong to $\mathcal{B}$.

Proof. Choose enumerations

$$G_0 = \{g_0^{(1)}, g_0^{(2)}, g_0^{(3)}, \dots\}, \qquad G_1 = \{g_1^{(1)}, g_1^{(2)}, g_1^{(3)}, \dots\},$$

of the elements in the countable groups G_0 and G_1 (with repetitions if G_0 or G_1 is finite). Let $G_i^{(n)}$ be the subgroup of G_i generated by $\{g_i^{(1)}, g_i^{(2)}, \dots, g_i^{(n)}\}$. We construct inductively C^*-algebras $A_1, A_2, A_3, \dots$ in $\mathcal{B}$, *-homomorphisms $\varphi_n \colon A_n \to A_{n+1}$, and group isomorphisms $\alpha_i^{(n)}$ making the diagram

(13.5)

$$
\begin{array}{ccccccc}
G_i^{(1)} & \overset{\iota}{\lhook\joinrel\longrightarrow} & G_i^{(2)} & \overset{\iota}{\lhook\joinrel\longrightarrow} & G_i^{(3)} & \lhook\joinrel\longrightarrow & \cdots \\[2pt]
{\scriptstyle \alpha_i^{(1)}}\big\uparrow & & {\scriptstyle \alpha_i^{(2)}}\big\uparrow & & {\scriptstyle \alpha_i^{(3)}}\big\uparrow & & \\[2pt]
K_i(A_1) & \underset{K_i(\varphi_1)}{\longrightarrow} & K_i(A_2) & \underset{K_i(\varphi_2)}{\longrightarrow} & K_i(A_3) & \longrightarrow & \cdots
\end{array}
$$

commutative for $i = 0, 1$. To obtain this, use first Corollary 13.1.3 to find A_1 and isomorphisms

$$\alpha_0^{(1)} \colon K_0(A_1) \to G_0^{(1)}, \qquad \alpha_1^{(1)} \colon K_1(A_1) \to G_1^{(1)}.$$

Use next Proposition 13.2.3 inductively, with

$$A = A_{n-1} \quad \text{and} \quad \alpha_i = \iota \circ \alpha_i^{(n-1)} \colon K_i(A_{n-1}) \to G_i^{(n)},$$

to find A_n in $\mathcal{B}$, a *-homomorphism $\varphi_{n-1} \colon A_{n-1} \to A_n$, and group isomorphisms $\alpha_i^{(n)} \colon K_i(A_n) \to G_i^{(n)}$, for $i = 0, 1$, making the diagram (13.5) commutative.

Let A be the inductive limit C^*-algebra of the sequence

$$A_1 \overset{\varphi_1}{\longrightarrow} A_2 \overset{\varphi_2}{\longrightarrow} A_3 \overset{\varphi_3}{\longrightarrow} \cdots .$$

By continuity of the functors K_0 and K_1 (see Theorem 6.3.2 and Proposition 8.2.7),

$$K_i(A) \cong \varinjlim (K_i(A_n), K_i(\varphi_n)) \cong \varinjlim (G_i^{(n)}, \iota) \cong \bigcup_{n=1}^{\infty} G_i^{(n)} = G_i$$

for $i = 0, 1$, and this proves the theorem. $\qquad\qquad\square$

13.3   Exercises

Exercise 13.1. Let n, k_0, k_1, d_0, d_1 be positive integers such that $n = d_0 k_0 = d_1 k_1$. Let $\varphi_j \colon M_{k_j}(\mathbb{C}) \to M_n(\mathbb{C})$ be the *-homomorphisms given by $\varphi_j(a) = \mathrm{diag}(a, a, \ldots, a)$, with d_j copies of a, for $j = 0, 1$. Let A be the sub-C^*-algebra of $C([0,1], M_n(\mathbb{C}))$ consisting of those functions f such that $f(0) = \varphi_0(a_0)$ and $f(1) = \varphi_1(a_1)$ for some a_0 in $M_{k_0}(\mathbb{C})$ and some a_1 in $M_{k_1}(\mathbb{C})$.

Construct a short exact sequence

$$0 \longrightarrow SM_n(\mathbb{C}) \longrightarrow A \longrightarrow M_{k_0}(\mathbb{C}) \oplus M_{k_1}(\mathbb{C}) \longrightarrow 0,$$

and use this sequence to calculate $K_0(A)$ and $K_1(A)$. Show that 0 and 1 are the only projections in A if and only if n is the least common multiple of d_0 and d_1.

Exercise 13.2. The purpose of this exercise is to realize each pair of finitely generated Abelian groups as the K-groups of an Abelian C^*-algebra.

(i) Let X be a compact Hausdorff space, and let x_0 be an element in X. Show that $C_0(X \setminus \{x_0\})^\sim$ is (isomorphic to) $C(X)$. Then use Exercise 12.2 to find for each integer $n \geqslant 2$ a locally compact Hausdorff space Y_n such that $K_0(C_0(Y_n))$ is isomorphic to $\mathbb{Z}/n\mathbb{Z}$, and $K_1(C_0(Y_n)) = 0$.

(ii) Show that for each pair of finitely generated Abelian groups G_0, G_1 there exists a locally compact Hausdorff space X such that $K_0(C_0(X)) \cong G_0$ and $K_1(C_0(X)) \cong G_1$. [Hint: Use (i), Example 10.2.5, and Exercise 4.2.]

Note. It can be shown that for every pair G_0, G_1 of countable Abelian groups there exists a locally compact Hausdorff space X such that $K_0(C_0(X)) \cong G_0$ and $K_1(C_0(X)) \cong G_1$.

Exercise 13.3. Let A and B be C^*-algebras in the class $\mathcal{B}$, and let

$$\alpha_0 \colon K_0(A) \to K_0(B) \quad \text{and} \quad \alpha_1 \colon K_1(A) \to K_1(B)$$

be group homomorphisms. By inspecting the proof of Proposition 13.2.3, show that there exist a natural number m and a *-homomorphism $\varphi \colon A \to M_m(B)$ such that $K_0(\varphi) = K_0(\lambda_m) \circ \alpha_0$ and $K_1(\varphi) = K_1(\lambda_m) \circ \alpha_1$.

This property of the class $\mathcal{B}$ is rare as the questions below will show.

(i) Let A and B be C^*-algebras, let $\alpha_0 \colon K_0(A) \to K_0(B)$ be a group homomorphism, and suppose that there exist a positive integer m and a $*$-homomorphism $\varphi\colon A \to M_m(B)$ such that $K_0(\varphi) = K_0(\lambda_m) \circ \alpha_0$. Show that $\alpha_0(K_0(A)^+)$ is contained in $K_0(B)^+$, i.e., α_0 is positive. [Hint: Look at Paragraph 5.1.8.]

(ii) Show that each C^*-algebra A in the class $\mathcal{B}$ is projectionless (its only projection is 0) and that $K_0(A)^+ = \{0\}$. Conclude that every group homomorphism $K_0(A) \to K_0(B)$ is positive when A and B belong to $\mathcal{B}$.

(iii) Let A and B be unital AF-algebras, and let $\alpha_0\colon K_0(A) \to K_0(B)$ be a group homomorphism. Show that there exist a positive integer m and a $*$-homomorphism $\varphi\colon A \to M_m(B)$ such that $K_0(\varphi) = K_0(\lambda_m) \circ \alpha_0$ if and only if α_0 is positive. [Hint: Look at Exercise 7.7.]

Give an example of AF-algebras A, B and a non-positive group homomorphism $\alpha_0\colon K_0(A) \to K_0(B)$.

(iv) As shown in Example 11.3.3, $K_0(C(S^2)) \cong \mathbb{Z} \oplus \mathbb{Z}$, and by Example 3.3.5 there is a surjective group homomorphism $\dim\colon K_0(C(S^2)) \to \mathbb{Z}$. It follows that the kernel of dim is isomorphic to $\mathbb{Z}$. Let b be a generator of this kernel. Let m be any positive integer, and put $B = M_m(\mathbb{C}) \oplus M_m(\mathbb{C})$. Show that $K_0(\varphi)(b) = 0$ for every $*$-homomorphism $\varphi\colon C(S^2) \to B$.

Find an example of a positive group homomorphism $\alpha_0\colon K_0(C(S^2)) \to K_0(\mathbb{C} \oplus \mathbb{C})$ for which there for no integer m exists a $*$-homomorphism $\varphi\colon C(S^2) \to M_m(\mathbb{C} \oplus \mathbb{C})$ with $K_0(\varphi) = K_0(\lambda_m) \circ \alpha_0$.

Exercise 13.4 (The C^*-algebra generated by two projections). Let A be the sub-C^*-algebra of $C([0,1], M_2(\mathbb{C}))$ consisting of all functions f such that

$$f(0) = \begin{pmatrix} d & 0 \\ 0 & 0 \end{pmatrix}, \qquad f(1) = \begin{pmatrix} d_1 & 0 \\ 0 & d_2 \end{pmatrix},$$

for some d, d_1, d_2 in $\mathbb{C}$, and consider the elements p, q in A given by

$$p(t) = \begin{pmatrix} 1 & 0 \\ 0 & 0 \end{pmatrix}, \qquad q(t) = \begin{pmatrix} 1-t & \sqrt{t(1-t)} \\ \sqrt{t(1-t)} & t \end{pmatrix}, \qquad t \in [0,1].$$

(i) Show that p and q are projections in A.

(ii) Show that $A = C^*(p,q)$. [Hint: By looking at the elements p, q, pq, and $(p-q)^2$ show that

(a) for all d, d_1, d_2 in $\mathbb{C}$ there exists f in $C^*(p,q)$ with $f(0) = \text{diag}(d,0)$ and $f(1) = \text{diag}(d_1, d_2)$,

(b) the function g, given by $g(t) = \text{diag}(t,t)$ for $t \in [0,1]$, belongs to $C^*(p,q)$,

(c) if $\{e_{i,j}\}_{i,j=1,2}$ are the standard matrix units for $M_2(\mathbb{C})$, and if $0 < \delta < 1/2$, then there exists $h_{i,j}$ in $C^*(p,q)$ with $\|h_{i,j}\| \leqslant 1$ and $h_{i,j}(t) = e_{i,j}$ for $t \in [\delta, 1-\delta]$.

Combine these results to prove (ii).]

(iii) Construct a short exact sequence

$$0 \longrightarrow SM_2(\mathbb{C}) \longrightarrow A \longrightarrow \mathbb{C} \oplus \mathbb{C} \oplus \mathbb{C} \longrightarrow 0\,,$$

and use this sequence to calculate $K_0(A)$ and $K_1(A)$. Show that

$$K_0(A) = \mathbb{Z}[p]_0 + \mathbb{Z}[q]_0.$$

(iv) Let D be a C^*-algebra, and let e, f be projections in D. Show that there is a (unique) *-homomorphism $\varphi \colon A \to D$ such that $\varphi(p) = e$ and $\varphi(q) = f$. Conclude that $C^*(e,f)$ is isomorphic to A/I for some closed two-sided ideal I in A.

[Hint: Define projections $e' = (e,p)$ and $f' = (f,q)$ in the C^*-algebra $D \oplus A$, and put $E = C^*(e', f')$. Let $\varphi_1 \colon E \to D$ and $\varphi_2 \colon E \to A$ be the restrictions of the coordinate mappings $D \oplus A \to D$ and $D \oplus A \to A$. It suffices to show that φ_2 is injective, since $\varphi = \varphi_1 \circ \varphi_2^{-1}$ then will be as desired (check this!).

Show that $z = (e' - f')^2$ is in the center of E, i.e., that $zx = xz$ for every x in E.

Show next that φ_2 is injective if for each irreducible representation $\pi \colon E \to B(H)$, on a Hilbert space H, there is a *-homomorphism $\psi \colon A \to B(H)$ such that $\pi = \psi \circ \varphi_1$. Let π be such an irreducible (non-degenerate) representation. Show that $\pi(z) = t \cdot 1$ for some t in the interval $[0,1]$. Describe the possible irreducible representations for $t = 0$ (here is one one-dimensional representation), for $t = 1$ (here are two one-dimensional representations), and for $0 < t < 1$ (here is — up to unitary equivalence — one representation for each t). Show that ψ exists in each case.]

References

[1] M. F. Atiyah, *K-theory*, 2nd ed., Advanced Book Program, Addison–Wesley, Redwood City, CA, 1967, 1989.

[2] B. Blackadar, A simple unital projectionless C^*-algebra, *J. Operator Theory* **5** (1981), 63–71.

[3] _____, *K-theory for operator algebras*, M. S. R. I. Monographs, vol. 5, Springer–Verlag, Berlin and New York, 1986.

[4] B. Blackadar and D. Handelman, Dimension functions and traces on C^*-algebras, *J. Funct. Anal.* **45** (1982), 297–340.

[5] B. Blackadar and M. Rørdam, Extending states on preordered semigroups and the existence of quasitraces on C^*-algebras, *J. Algebra* **152** (1992), 240–247.

[6] R. Bott, Homotopy of classical groups, *Ann. Math.* **70** (1959), no. 2, 313–337.

[7] O. Bratteli, Inductive limits of finite dimensional C^*-algebras, *Trans. Amer. Math. Soc.* **171** (1972), 195–234.

[8] L. G. Brown, R. Douglas, and P. Fillmore, Unitary equivalence modulo the compact operators and extensions of C^*-algebras, *Proceedings of a conference on operator theory, Halifax, Nova Scotia*, Lecture notes in Math., vol. 345, Springer–Verlag (Berlin, Heidelberg and New York), 1973, pp. 58–128.

[9] _____, Extensions of C^*-algebras and K-homology, *Ann. Math.* **105** (1977), 265–324.

[10] L. Coburn, The C^*-algebra of an isometry, *Bull. Amer. Math. Soc.* **73** (1967), 722–726.

[11] A. Connes, *Noncommutative geometry*, Academic Press, San Diego, CA, 1994.

[12] J. Cuntz, Simple C^*-algebras generated by isometries, *Comm. Math. Phys.* **57** (1977), 173–185.

[13] _____, K-theory for certain C^*-algebras, *Ann. Math.* **113** (1981), 181–197.

[14] H. G. Dales, *Banach algebras and automatic continuity*, Oxford University Press, 2000.

[15] K. R. Davidson, C^*-*algebras by example*, Fields Institute Monographs, Amer. Math. Soc., Providence, R.I., 1996.

[16] E. G. Effros, *Dimensions and C^*-algebras*, CBMS Regional Conference Series in Mathematics, vol. 46, Amer. Math. Soc., Washington, D.C., 1981.

[17] E. G. Effros, D. E. Handelman, and C.-L. Shen, Dimension groups and their affine representations, *Amer. J. Math.* **102** (1980), 385–407.

[18] G. A. Elliott, On the classification of inductive limits of sequences of semisimple finite-dimensional algebras, *J. Algebra* **38** (1976), 29–44.

[19] _____, The classification problem for amenable C^*-algebras, *Proceedings of the International Congress of Mathematicians*, vol. 1,2, Birkhäuser, Basel, 1995, pp. 922–932.

[20] P. A. Fillmore, *A user's guide to operator algebras*, Canadian Mathematical Society Series of Monographs and Advanced Texts, John Wiley & Sons, New York, 1996.

[21] J. Glimm, On a certain class of operator algebras, *Trans. Amer. Math. Soc.* **95** (1960), 318–340.

[22] K. R. Goodearl, *Partially ordered abelian groups with interpolation*, Mathematical Surveys and Monographs, no. 20, Amer. Math. Soc., Providence, R.I., 1986.

[23] U. Haagerup, Every quasi-trace on an exact C^*-algebra is a trace, preprint, Odense University, 1991.

[24] G. Higman, The units of group-rings, *Proc. London Math. Soc.* **46** (1940), no. 2, 231–248.

[25] D. Husemoller, *Fibre bundles*, 3rd ed., Graduate Texts in Mathematics, no. 20, Springer Verlag, New York, 1966, 1994.

[26] R. V. Kadison and J. R. Ringrose, *Fundamentals of the theory of operator algebras*, Academic Press, London, 1986.

[27] G. G. Kasparov, The operator K-functor and extensions of C^*-algebras, *Math. USSR–Izv.* **16** (1981), 513–672, English translation.

[28] E. Kirchberg, The classification of purely infinite C^*-algebras using Kasparov's theory, to appear in the *Fields Institute Communication series*.

[29] G. J. Murphy, *C^*-algebras and operator theory*, Academic Press, London, 1990.

[30] G. K. Pedersen, *C^*-algebras and their automorphism groups*, Academic Press, London, 1979.

[31] _____, *Analysis now*, Graduate Texts in Mathematics, no. 120, Springer-Verlag, New York-Berlin, 1988.

[32] N. C. Phillips, A classification theorem for nuclear purely infinite simple C^*-algebras, *Documenta Math.* (2000), no. 5, 49–114.

[33] M. Pimsner and D. V. Voiculescu, Imbedding the irrational rotation algebras into an AF-algebra, *J. Operator Theory* **4** (1980), 201–210.

[34] _____, K-groups of reduced crossed products by free groups, *J. Operator Theory* **8** (1982), 131–156.

[35] R. T. Powers, Simplicity of the C^*-algebra of the free group on two generators, *Duke J. Math.* **42** (1975), 151–156.

[36] M. Rieffel, C^*-algebras associated with irrational rotations, *Pacific J. Math.* **93** (1981), 415–429.

[37] M. A. Rieffel, Dimension and stable rank in the K-theory of C^*-algebras, *Proc. London Math. Soc.* **46** (1983), no. (3), 301–333.

[38] W. Rudin, *Functional analysis*, Tata McGraw-Hill, New York, 1966.

[39] R. G. Swan, Vector bundles and projective modules, *Trans. Amer. Math. Soc.* **105** (1962), 264–277.

[40] N. E. Wegge-Olsen, *K-theory and C^*-algebras*, Oxford University Press, New York, 1993.

Table of K-groups

C^*-algebra A	$K_0(A)$	$K_1(A)$	References
$\mathbb{C}$, $M_n(\mathbb{C})$	$\mathbb{Z}$	0	Example 3.3.2, Example 8.1.8.
$\mathcal{K}$	$\mathbb{Z}$	0	Corollary 6.4.2, Example 8.2.9.
$B(H)$	0	0	Example 3.3.3, Example 8.1.8.
$\mathcal{Q}(H)$	0	$\mathbb{Z}$	Example 9.4.3, Exercise 12.4.
$C([0,1])$, $C(\mathbb{D})$	$\mathbb{Z}$	0	Exercise 3.3, Exercise 8.2.
$C(X)$, X contractible	$\mathbb{Z}$	0	Example 3.3.6, Exercise 8.2.
$C_0((0,1])$	0	0	Exercise 4.2, Exercise 8.4.
$C_0((0,1))$, $C_0(\mathbb{R})$	0	$\mathbb{Z}$	Example 11.3.2.
$C(S^1)$, $C(\mathbb{T})$	$\mathbb{Z}$	$\mathbb{Z}$	Example 11.3.3.
$C(S^n)$, n even	$\mathbb{Z} \oplus \mathbb{Z}$	0	Example 11.3.3.
$C(S^n)$, n odd	$\mathbb{Z}$	$\mathbb{Z}$	Example 11.3.3.
$C_0(\mathbb{R}^n)$, n even	$\mathbb{Z}$	0	Example 11.3.2.
$C_0(\mathbb{R}^n)$, n odd	0	$\mathbb{Z}$	Example 11.3.2.
$C(\mathbb{T}^n)$	$\mathbb{Z}^{2^{n-1}}$	$\mathbb{Z}^{2^{n-1}}$	Example 11.3.4.
$C(\mathbb{D}/\sim_n)$, Moore space	$\mathbb{Z} \oplus (\mathbb{Z}/n\mathbb{Z})$	0	Exercise 12.2.
$C(Z_n)$, n-clover	$\mathbb{Z}$	$\mathbb{Z}^n$	Exercise 12.3.

C^*-algebra A	$K_0(A)$	$K_1(A)$	References
A_θ, irrational rotation algebra	$\mathbb{Z} \oplus \mathbb{Z}$	$\mathbb{Z} \oplus \mathbb{Z}$	Exercise 5.8.
$\mathcal{O}_n$, Cuntz algebra	$\mathbb{Z}/(n-1)\mathbb{Z}$	0	Exercise 4.5, Exercise 8.10.
$\mathcal{T}$, Toeplitz algebra	$\mathbb{Z}$	0	Example 9.4.4, Exercise 12.4.
$\mathcal{T}_0$, reduced Toeplitz algebra	0	0	Example 9.4.4, Exercise 12.4.
D_n, dimension drop algebra	0	$\mathbb{Z}/n\mathbb{Z}$	Paragraph 13.1.1.
$M_n(B)$	$K_0(B)$	$K_1(B)$	Proposition 4.3.8, Proposition 8.2.8.
$\mathcal{K}B$, stabilization	$K_0(B)$	$K_1(B)$	Proposition 6.4.1, Proposition 8.2.8.
$\widetilde{B}$, unitization	$K_0(B) \oplus \mathbb{Z}$	$K_1(B)$	Example 4.3.5, equation (8.4).
$B_1 \oplus B_2$	$K_0(B_1) \oplus K_0(B_2)$	$K_1(B_1) \oplus K_1(B_2)$	Proposition 4.3.4, Proposition 8.2.6.
SB, suspension	$K_1(B)$	$K_0(B)$	Theorem 10.1.3, Corollary 11.3.1.
CB, cone	0	0	Example 4.1.5, Exercise 8.3.
$\mathbb{T}B$	$K_0(B) \oplus K_1(B)$	$K_0(B) \oplus K_1(B)$	Example 11.3.4.
AF-algebra	Dimension group	0	Proposition 7.2.8, Exercise 8.7.
UHF-algebra, type n	$Q(n)$	0	Theorem 7.4.5, Exercise 8.3.
II_1-factor	$\mathbb{R}$	0	Exercise 3.12, Exercise 8.14.
Inductive limit of dimension drop algebras	arbitrary countable	arbitrary countable	Theorem 13.2.4.

Index of symbols

General index

Printed in the United States
By Bookmasters